PHILOSOPHICAL PROBLEMS OF MANY-VALUED LOGIC

SYNTHESE LIBRARY

A SERIES OF MONOGRAPHS ON THE

RECENT DEVELOPMENT OF SYMBOLIC LOGIC,

SIGNIFICS, SOCIOLOGY OF LANGUAGE,

SOCIOLOGY OF SCIENCE AND OF KNOWLEDGE

STATISTICS OF LANGUAGE

AND RELATED FIELDS

Editors:

B. H. KAZEMIER / D. VUYSJE

A. A. ZINOV'EV

PHILOSOPHICAL PROBLEMS OF MANY-VALUED LOGIC

A revised edition,

edited and translated by Guido Küng

and David Dinsmore Comey

D. REIDEL PUBLISHING COMPANY / DORDRECHT-HOLLAND

1963

Printed in The Netherlands by Bosch, Utrecht

CONTENTS

FOREWORD FROM THE EDITORS

This work, written by a young Russian logician at the Institute of philosophy in Moscow, fills a long-empty gap in our literature about many-valued logics. The only previous monograph on many-valued logic in English is primarily mathematical and technical, and only in its introduction does it treat the philosophical problems raised by many-valued logic.[1)]

In the present work, the author takes up such questions as the 'unity' and 'diversity' of logic as a whole, the empirical relationship of logic to its object domains, and the relationship between ordinary thinking on the one hand and two-valued and many-valued logics on the other. He also treats briefly the relationship between many-valued logic and the criticisms raised by dialectical logic against traditional formal logic, a section which despite its brevity will be useful to those who are interested in Soviet philosophy.

Philosophers will also find that the author introduces his technical symbolic treatment in gradual stages, so that the book should present no technical difficulty to anyone who has had the usual undergraduate course in symbolic logic. A number of alternative systems of many-valued logics are presented, and their relationship to ordinary two-valued logic is discussed at considerable length. The book can thus serve as an introductory work for those who wish to acquire a working knowledge of the calculi of many-valued logical systems.

It should perhaps be made clear that this is *not* a translation of Zinov'ev's *Filosofskie Problemy Mnogoznačnoj Logiki* (Philosophical Problems of Many-valued Logic) published in 1960.[2)] Instead, the author was kind enough to prepare a considerably revised version of his earlier work especially for this English-language edition. The entire fourth chapter on

1) J. B. Rosser and A. R. Turquette, *Many-valued Logics*, North-Holland Publishing Co., Amsterdam, 1952.

2) Institut Filosofii, Izdatel'stvo Akademii Nauk SSSR, Moscow, 1960, 139 pp. Reviewed by Comey in *The Journal of Symbolic Logic*, and by J. M. Bocheński and L. H. Hackstaff in 'A Study in Many-valued Logic', *Studies in Soviet Thought*, Vol. 2, 1962, 37-48; corrections by Comey of the latter review appeared in *Studies in Soviet Thought*, Vol. 2, 1962, 241-242.

two-valued and many-valued logics is new, and so many other sections of the book have been supplemented with new material that it would be impossible to list here the differences between this revised edition and the earlier Russian volume.

The editors have changed the order of presentation of the various alternative axiomatic systems in the second chapter, with the approval of the author. Otherwise, the editors have confined their own revisions to technical changes in the logical formulae and manner of presentation which they hope will make certain points clearer and more rigorous. They have provided separate name and subject indexes for this edition. The translation of the work is a fairly literal one; it was not necessary to make any changes in order to match the author's logical terminology with that of contemporary Western usage.

There is no necessity for going into a detailed description here of the contemporary state of logical research in the Soviet Union, since such accounts are now available elsewhere.[3]) However, it should be noted that most of the research in symbolic logic in the Soviet Union has been done by mathematicians, and that Zinov'ev is the only Russian philosopher writing on symbolic logic who has a first-hand technical knowledge of modern mathematical logic. A complete list of his published works is given following this Foreword.

The author was born in 1922, and became a university undergraduate at the age of sixteen. His studies were interrupted by the Second World War, during which he was a pilot of a *šturmovik* (low-level attack plane) and saw action at the front in eastern Europe. He was married in 1951, and has one daughter. He received his *kandidat* degree[4]) in 1954, and his dissertation on the relationship between the abstract and the concrete [1] was highly praised by the scientists attending the oral defense. He now works as logician at the Institute of Philosophy of the Academy of Sciences of the USSR in Moscow, and has organized the publication of three sub-

3) Cf. J. M. Bocheński, 'Soviet Logic', *Studies in Soviet Thought*, Vol. 1, 1961, 29-38; G. Küng, 'Mathematical Logic in the Soviet Union (1917-1947 and 1947-1957)', *Ibid.*, 39-43; G. Küng, 'Bibliography of Soviet Work in the Field of Mathematical Logic and the Foundations of Mathematics, from 1917-1957' (with a preface by J. M. Bocheński), *The Notre Dame Journal of Formal Logic*, Vol. 3, 1962, 1-40; and D. D. Comey, 'Two Recent Soviet Conferences on Logic', *Studies in Soviet Thought*, Vol. 2, 1962, 21-36.

4) The *kandidat* degree in the Soviet Union roughly corresponds to an American Ph. D.

stantial *sborniki* (collections of articles) on logic published by the Institute of Philosophy [14, 19, 25]. In addition, he has written two books [22, 24] and a score of articles. These works cover roughly five fields: (I) the logic of connections (or in Reichenbach's terminology, 'physical implication') [12, 14, 16, 17, 19, 20]; (II) scientific methodology [1, 6, 8, 13, 15, 18]; (III) logical implication [24]; (IV) various questions on logic and epistemology [3, 7, 9, 10, 21]; and (V) problems of many-valued logic [11, 22, 23, 25].[5)]

GUIDO KÜNG

DAVID DINSMORE COMEY

5) The reader can find reviews by Comey of most of these works in *The Journal of Symbolic Logic*.

BIBLIOGRAPHY OF THE PUBLICATIONS OF ALEKSANDR ALEKSANDROVIČ ZINOV'EV

(Listed in chronological order; all works cited are in Russian except for [6, 8] which are in Czech, and [9] which was published in Polish. For an explanation of the system of transliteration from the Russian, see the transliteration table on page 149.)

[1] *Metod vosxoždenija ot abstraktnogo k konkretnomu* (The method of ascending from the abstract to the concrete): avtoreferat dissertacii (dissertation abstract), Moscow, 1954.

[2] Rasširjat' tematiku logičeskix issledovanij (Broadening the subject matter of logical investigations), *Voprosy Filosofii*, 1957, 3, 211-215.

[3] O razrabotke dialektiki kak logiki (On the development of dialectic as logic), *Voprosy Filosofii*, 1957, 4, 188-190.

[4] O knige 'Myšlenie i jazyk' (On the book 'Thought and Language'), *Voprosy Jazykoznanija*, 1958, 2, 141-145.

[5] O rabote seminara po logike v Institute Filosofii AN SSSR (On the work of the seminar on logic at the Institute of Philosophy of the Academy of Sciences of the USSR), *Voprosy Filosofii*, 1958, 2, 167-172.

[6] K problému abstraktního a konkrétního poznatku (On the problem of abstract and concrete knowledge), *Filosofický Časopis*, 1958, 2.

[7] O logike normativnyx predloženij (On the logic of normative sentences), *Voprosy Filosofii*, 1958, 11, 156-159.

[8] K otázce hodnocení soudů jako abstraktních a konkrétních (On the question of evaluating judgments as abstract and concrete), *Filosofický Časopis*, 1959, 1.

[9] O logicznej niesprzeczności sądów prawdziwych dotyczących zmian i związków między przedmiotami (On the logical non-contradiction of true statements concerning changes and connections between objects), *Studia filozoficzne*, 1959, 1, 124-133; translated into German by G. Küng as 'Über die logische Widerspruchsfreiheit der wahren Urteile über Veränderungen und Zusammenhänge unter Gegenständen', in: Nikolaus Lobkowicz, *Das Widerspruchsprinzip in der neueren sowjetischen Philosophie*, (Sovietica, Veröffentlichungen des Ost-Europa Instituts, Universität Freiburg/Schweiz). D. Reidel Publishing Co., Dordrecht, Holland, 1959, pp. 65-76.

[10] Ob odnoj programme issledovanija myšlenija (On one program of investigating thought), *Doklady Akademii Pedagogičeskix Nauk RSFSR*, 1959, 2, 71-74.

[11] Problema značenij istinnosti v mnoggoznačnoj logike (The problem of truth-values in many-valued logic), *Voprosy Filosofii*, 1959, 3, 131-136.

[12] Sledovanie kak svojstvo vyskazyvanij o svjazjax (Implication as a property of propositions concerning connections), *Naučnye Doklady Vysšej Školy: Filosofskie Nauki*, 1959, 3, 84-88.

[13] O matematičeskoj lingvistike (On mathematical linguistics), *Voprosy Filosofii*, 1959, 9, 133-140.

[14] Logičeskoe stroenie znanij o svjazjax (The logical structure of knowledge concerning connections), *Logičeskie Issledovanija: sbornik statej* (Logical Investigations: a collection of articles), Institut Filosofii, Izdatel'stvo Akademii Nauk SSSR, Moscow, 1959, pp. 113-138.

[15] Logičeskaja model' kak sredstvo naučnogo issledovanija (The logical model as a means of scientific research) (Co-author: I. I. Revzin), *Voprosy Filosofii*, 1960, 1, 82-90.

[16] K opredeleniju ponjatija svjazi (On the definition of the concept of connection), *Voprosy Filosofii*, 1960, 8, 58-66.

[17] K voprosu o metode issledovanija znanij (vyskazyvanija o svjazjax) (On the question of a method of investigating knowledge (propositions about connections)), *Doklady Akademii Pedagogičeskix Nauk RSFSR*, 1960, 3.

[18] O logičeskoj prirode vosxoždenija ot abstraktnogo k konkretnomu (On the logical nature of ascending from the abstract to the concrete), *Filosofskaja Énciklopedija*, Tom I, Institut Filosofii, Izdatel'stvo 'Sovetskaja Énciklopedija', Moscow, 1960, pp. 295-298.

[19] Deduktivnyj metod v issledovanii vyskazyvanij o svjazjax (The deductive method in the investigation of propositions concerning connections), *Primenenie Logiki v Nauke i Texnike: sbornik statej* (Applications of Logic to Science and Technology: a collection of articles), Institut Filosofii, Izdatel'stvo Akademii Nauk SSSR, Moscow, 1960, pp. 215-242.

[20] K voprosu ob obščnosti vyskazyvanij o svjazjax (On the question of the generality of propositions about connections), *Ibid.* pp. 243-250.

[21] Ob odnom variante teorii opredelenij (On one variant of the theory of definitions), *Ibid.* pp. 251-262.

[22] *Filosofskie problemy mnogoznačnoj logiki* (Philosophical problems of many-valued logic), Institut Filosofii, Izdatel'stvo Akademii Nauk SSSR, Moscow, 1960, 139 pp.

[23] Ob odnom sposobe obzora funkcij istinnosti *n*-značnogo isčislenija vyskazyvanij (On one means of viewing truth functions of an *n*-valued propositional calculus), *Studia Logica*, 1961, 11, 217 f.

[24] *Logika vyskazyvanij i teorija vyvoda* (The logic of propositions and the theory of deduction), Institut Filosofii, Izdatel'stvo Akademii Nauk SSSR, Moscow, 1961, 152 pp.

[25] Dvuznačnaja i mnogoznačnaja logika (Two-valued and many-valued logic), *Filosofskie Voprosy Sovremennoj Formal'noj Logiki:* sbornik statej (Philosophical Problems of Contemporary Formal Logic: a collection of articles), Institut Filosofii, Izdatel'stvo Akademii Nauk SSSR, Moscow, 1962, pp. 111-139.

AUTHOR'S PREFACE

This book is a revised version of ZINOV'EV 1960. The changes made involve not only the form of the exposition, but also affect the very substance of the work.

In connection with the origin and the development of many-valued logic several interesting philosophical problems have been raised, the consideration of which is still in its beginning stages, and in any case has been undertaken only in very insignificant dimensions, somewhere at the periphery of philosophy. We will here examine some of these problems, without any pretension of completely exhausting their scope or categorically giving their solution. Thus we will consider these problems in their simplest form because in our opinion it is possible in this way to successfully achieve elementary philosophical clarity. In considering the sequence of analysis, one has to take into account the following circumstances (over and above the subjective shortcomings of the author): there is no necessary (logically determined) order of exposition of the detailed philosophical problems of logic in this work, because this is not a systematic exposition of logic or of philosophy; therefore it will occasionally be necessary to make digressions, to return to problems considered earlier, to give enumerations which do not fulfil the strict rules of division, etc. Such shortcomings are unavoidable if 'descriptive' methods are to play any rôle at all in logic.

The task of this book is to give an exposition of the philosophical interpretation of those problems of many-valued logic which seem important to us, and to give them an interpretation which seems more correct to us. We mention this only so that the reader will not expect an historical outline and encyclopaedic information about the development of many-valued logic and its philosophical interpretation. Also, compared with the works of Łukasiewicz, who was not only a founder of many-valued logic (together with Post) but also the founder of a philosophical interpretation of the problems connected with it, this book has an entirely different outlook, for it mainly considers other problems.

The term 'philosophical problems' is, of course, not so well defined that it could be accepted as absolute and universal. But we will not elaborate on this subject. We simply use this term as a heading under which the problems considered below are grouped. If anyone rejects the use of the term 'philosophical problems' in this connection, we will not object to his substituting for it any other more convenient and non-controversial term. In any case such problems are usually considered as philosophical (in the good sense of the word).

The consideration of philosophical questions of logic has to meet the difficulty that an important part of logic (at least because of the historical origin of its ideas, terms, etc.) appears as a philosophical discipline (solving problems which for centuries have been considered as philosophical). In this work we will endeavour to consider the philosophical questions of many-valued logic in the same spirit and on the same level on which the philosophical questions of other concrete (non-philosophical) sciences are considered, and if possible to escape the afore-mentioned difficulty. To consider the questions of logic which are especially concerned with the historical peculiarity indicated would require a much more extensive work.

In this book we mainly consider questions connected with the logic of propositions. We will only treat the logic of predicates briefly. This is necessary because one has to become familiar with the philosophical problems of many-valued logic on the level of the logic of propositions before attacking the logic of predicates. Besides, the philosophical questions considered in this work appear to be the same for both. The many-valuedness of propositions (the admission of more than two truth values) in no way touches their internal structure. The problems of many-valued predicate logic are therefore more a question of logical technique than of philosophy. In any case, what is philosophical in them can to a great extent be reduced to the philosophical questions of the logic of propositions (e.g. by considering the quantifiers as a kind of abbreviation.)

In many cases we will have to touch upon general philosophical questions of logic. But we will endeavour to reduce such considerations to the necessary minimum and give our primary attention to what refers specifically to many-valued logic and to its influence on the statement of problems of a more general order.

The philosophical problems of many-valued logic can more or less be divided into two groups. To the first group belong the problems which simply concern propositions, truth, falsity, etc., or in general, arguments, functions, values, etc. To the second group belong those problems whose solution is not possible except by an analysis of the concepts 'proposition', 'true', 'false', etc. In some sense the second kind of problems seems more philosophical. But we will consider them later, and connect them with one of the interpretations of logical constructions.

The present work is aimed at the relatively wide circle of the scientific public which is interested in the philosophical and logical (logico-philosophical) problems of contemporary science. Therefore we will first give some examples of many-valued logical constructions, the analysis of which will provide us with a sufficient base for our general considerations. In referring to different logical systems we will, of course, consider only their principles, and pick out various fragments, statements, and properties. Purely technical questions (i.e. of logical technique or mathematics) we will either leave out entirely, or present in an abbreviated form as illustrations of general ideas and methods of logic.

CHAPTER ONE

THE CONCEPT OF MANY-VALUED LOGIC

§ 1. Variable[1]) propositions[2]) will be designated by small Latin letters, and operators[3]), which convert them into new propositions and form complex propositions out of them, will be designated by large Latin letters written in front of the symbols for propositions: $N(x)$, $F^1(x)$, $A(x, y, z)$, $F^2(x^1,\ldots, x^m)$, etc. $N(x)$ is here a proposition the truth value of which depends in a definite way on (is a function of) the truth value x, and N indicates the type (the character, the form) of this dependence; in other words, $N(x)$ is a truth function (or a propositional function) of the type N of x. Similarly $F^1(x)$ is a truth function of the type F^1 of x; $A(x, y, z)$ is a truth function of the type A of x, y and z; $F^2(x^1,\ldots, x^m)$ is a truth function of the type F^2 of $x^1,\ldots, x^m$, etc. In place of propositional variables one can substitute propositions made up of propositional variables and the operators N, A, F^1, F^2, etc. (i.e., propositional functions).

In order not to complicate the exposition, we will not distinguish between truth functions and propositional functions, as has been done in, e.g., ROSSER and TURQUETTE: when speaking of variable propositions, we will mean any propositions whatsoever, and only take into account that they have one out of a given set of truth values. (It is for this reason that the mathematical term 'variable' has been introduced into logic). Because the meaning of the propositions is irrelevant to the determination of the types of relations according to which the truth values of compound propositions depend on the truth values of their component propositions, these relations can actually be regarded as truth functions – as functions in the strict sense of the word.

In many cases (if not in most) we will deal with functions of one or of two arguments. In these cases it will be useful to simplify the notation by making use of a parenthesis-free notation (the so-called Polish notation). The basic functions of propositional logic will have the following form in this notation:

Nx – negation (not-x),

Kxy – conjunction (x and y),

$\dot{A}xy$ – weak or non-exclusive disjunction (x or y; at least one of x and y),

Exy – strong or exclusive disjunction (either x or y; one and only one of x and y),[4])
Cxy – material implication (if x then y; not-x or y),
Rxy – equivalence (x if and only if y),[4])

In reading complex expressions one has to observe the following rules:

1) the symbol N refers only to the single proposition which follows immediately to its right;

2) the other operators K, A, E, C, R, etc., refer to exactly two propositions following immediately to their right.

For example, in *KNRxAxNyz* the second operator N refers only to y, A connects x and Ny, R connects x and *AxNy*, the first operator N refers to *RxAxNy*, K connects *NRxAxNy* and z. Filling in parentheses we get $K\{N[Rx(Ax\{Ny\})]\}z$. The number of parentheses must still be greater, if a notation of the type $F(x^1, \ldots, x^m)$ is used. For example, *RxAxNy* would take the form R[x, A(x, N(y))].

In some cases, however, the parenthesis-free notation is not convenient. For instance, the exclusive disjunction of three arguments $E(x, y, z)$ cannot be formulated in parenthesis-free notation using only the operator E, because $E(x, y, z)$, *EExyz* and *EExzy* are not equivalent. And formulated with the help of the operators A, K and N this disjunction takes the cumbersome form *AAKKxNyNzKKNxyNzKKNxNyz*.[5]) In such cases it is better to use the notation explained in the beginning of this section.

Our choice of the functions K, A, E, N and C as examples is not accidental. The motivation lies in the fact that, first of all, these functions correspond to the logical words 'and', 'or', 'not', 'if... then...' of ordinary language, and that secondly most logical systems contain them.

For defining propositional functions and describing their properties and relations, we will use (besides the truth table notation to which we will have recourse only in exceptional cases) the signs of equality, inequality, addition and subtraction. An equation of the form $a = \alpha$ will signify:

1) that the truth value of proposition a is equal to (identical with) the truth value of proposition α; or

2) that the truth value of the proposition a is equal to α, where α is any number corresponding to one of the truth values or any algebraic sum of truth values; for example, $a = n - b + c$ shall signify that the truth value of proposition a is equal to number n minus the truth value of

proposition *b* (some number), plus the truth value of proposition *c* (also some number). The signs for 'smaller' and 'larger' will be used similarly.

The signs of equality, inequality, addition and subtraction are, of course, not symbols of the logical systems under consideration, but symbols belonging to the language by means of which we describe the properties and mutual relations of the propositional functions. In operating with them, therefore, we should make use of supplementary symbols in order to indicate that we do not equate, add and subtract propositions themselves, but only their truth values. But this would unduly complicate the notation because, once we have agreed to use these and have agreed on the meaning of the corresponding expressions, recourse to a more complicated form is no longer necessary. The equality sign has an immediate informal meaning: it signifies the identification of truth values whereby the latter may be denoted by differently formulated expressions. There are no mathematical signs having such an immediate informal interpretation, which could be adopted when the truth values were designated by simple words ('true', 'false', 'undetermined') or by any chance symbols. Our notation implies that we have ordered the truth values in a definite way, corresponding, e.g., to the numbers *1*, *2*, *3*,... (i.e. it implies that we have numbered them) and this gives us the possibility of investigating the truth functions systematically.

With respect to the equality sign the following must be noted. In two-valued logic the equality sign coincides with *R*; i.e., $x = y$ is equivalent to *Rxy*. This circumstance makes it possible to replace many sentences about formulas of logic by formulas of this logic itself. For instance, instead of $NNx = x$, $Cxy = ANxy$, $NKxy = ANxNy$, etc., we can write *RNNxx*, *RCxyANxy*, *RNKxyANxNy*, etc. In many-valued logic, however, such a replacement is not always suitable nor permissable, and a strict distinction between symbols of the system under construction and the symbols of the language used to describe this construction (the metalanguage) is necessary.

The symbolization of the truth values will be explained in every case where it is necessary. In order to avoid too great a number of symbols, and for convenience in writing equations, we will often use the same symbols *K*, *A*, *N*, *C*, etc., with somewhat different meanings. This will not lead to confusion, as we will always make a corresponding convention. Instead of the terms 'proposition', 'truth value', etc., we could have adopted purely mathematical terms like 'formula', 'argument', 'value', etc. But this would not have made any essential difference and the logical terminology seems to be more customary and appropriate to ... logic.

§ 2. We will come back more than once to the question of two-valued logic. We will limit ourselves here to the following preliminary remarks.

By two-valued logic we first of all mean logical systems in which the propositions take one of two possible truth values (logical values), normally designated by the terms 'true' and 'false' (or correspondingly by the symbols *1* and *0*, *1* and *2*, *t* and *f*, etc.). In other words, by two-valued logic we first of all mean logical theories which are based on the following presupposition: the set of all propositions is divided exhaustively into two non-intersecting subsets. One of these subsets corresponds to truth (is the set of all true propositions), the other to falsity (is the set of all false propositions); thus no true proposition can be false and no false one true (cf. GRENIEWSKI 1957). We have intentionally said 'first of all', because as we shall see below, this is not the whole meaning of the term 'two-valued logic'.

The most basic logical system in which the afore-mentioned presupposition of the two-valuedness of propositions is initially assumed, is the construction of propositional logic with the help of two-valued matrices, or the two-valued algebra of propositions (cf. NOVIKOV). A detailed exposition of the system can be found in any text book or monograph in a more or less extensive form (cf.; e.g.; HILBERT and ACKERMANN, KLEENE, TARSKI, NOVIKOV, GRENIEWSKI 1955, MOSTOWSKI). We limit ourselves to brief considerations which we shall need for later reference.

Let number *1* designate truth and number *0* falsity. Then the functions of two-valued logic as introduced in § 1 can be defined as follows:

1) $Kxy = min(x, y)$, i.e., the truth value of Kxy is equal to the smaller of the truth values x and y;

2) $Axy = max(x, y)$, i.e., the truth value of Axy is equal to the larger of the truth values x and y;

3) $Cxy = max(Nx, y)$; or also: $Cxy = 1$ if $x \leqslant y$, and $Cxy = 0$ if $x > y$;

4) $Exy = 1$ if $x > y$ or $y > x$, and $Exy = 0$ if $x = y$; or: $Exy = $ $= min\,[max\,(x, y), max\,(1—x, 1—y)]$;

5) $Nx = 1 - x$; or: $Nx = 1$ if $x = 0$, and $Nx = 0$ if $x = 1$;

6) $Rxy = 1$ if $x = y$; $Rxy = 0$ if $x > y$ or $y > x$. Propositions which always have value *1* (for any truth values of their arguments), are called 'tautologies', 'always true propositions' ,'laws', etc. Here are some of the laws of the two-valued logic of propositions:

Rxx – law of identity,
AxNx – law of the excluded middle,
NKxNx – law of contradiction,[6])
RNKxyANxNy and *RNAxyKNxNy* – laws of De Morgan,
RKxAyzAKxyKxz and *RAxKyzKAxyAxz* – laws of distributivity,
RCNxyCNyx – law of contraposition,
RRxyKCxyCyx – law of the definition of equivalence,
RCxyANxy – law of the definition of implication,
CCxNxNx – *reductio ad absurdum.*

It is not necessary to give a more detailed exposition inasmuch as this is sufficiently well done in many logic courses. We can refer, say, to Hilbert and Ackermann, where a good characterization of the different ways of constructing two-valued propositional logic can be found.

We will now consider the law of the excluded middle and the law of contradiction. The reason for dealing separately with just these laws is clear: they stand in the center not only of the discussion concerning the relation between formal and non-formal (in Soviet philosophy: dialectical) logic, but also of the discussion concerning the relation between classical and non-classical logic within the limits of formal logic. These laws express the principle of two-valuedness even without explicit reference to the truth values ('*x* or not-*x*' and 'it is never the case that *x* and not-*x*'). They coincide very closely with the corresponding laws of traditional logic, so that they can be considered as explications of the latter.

The law of the excluded middle should, according to traditional logic, be rendered as *ExNx* ('either *x* or not-*x*'). But since *RExNxAxNx*, or $ExNx = AxNx$, holds, and since the function *A* is more convenient than *E* for several reasons, *AxNx* is chosen as the law of the excluded middle.

The notation adopted here, using equations, is not only more economical than the use of truth tables, but it also serves as a means for proving theorems. We will give some examples. Take the assertion $NKxy =$

$ANxNy$. According to the definitions $1\text{–}min(x, y) = max(1\text{–}x, 1\text{–}y)$. If $x = y$, then $1\text{–}min(x, y) = 1\text{–}x = 1\text{–}y$ and $max(1\text{–}x, 1\text{–}y) = 1\text{–}x = 1\text{–}y$, i.e., the assertion is true. If $x > y$, then $1\text{–}min(x, y) = 1\text{–}y$ and $max(1\text{–}x, 1\text{–}y) = 1\text{–}y$, i.e., the assertion is also true. Similarly if $x < y$, we get $1\text{–}x = 1\text{–}x$. Or take $NNx = x$: according to the definition we have $NNx = 1\text{–}Nx = 1\text{–}1 + x = x$. $NKxNx = 1\text{–}KxNx = 1\text{–}min(x, 1\text{–}x) = 1$, because either x or $1\text{–}x$ equals zero. $CCxNxNx = max(NCxNx, Nx) = max(1\text{–}CxNx, 1\text{–}x) = max(1\text{–}max(Nx, Nx), 1\text{–}x) = max(1\text{–}1 + x, 1\text{–}x) = max(x, 1\text{–}x) = 1$, because either x or $1\text{–}x$ equals 1.

The form of the definitions of the functions will vary depending on the choice of the numbers designating the truth values. Thus, if by 1 we designate truth and by 2 we designate falsity, then $Nx = 2\text{–}x + 1$, $Kxy = max(x, y)$, $Axy = min(x, y)$, $Cxy = min(3\text{–}x, y)$. But this does not affect the laws of logic. For instance, in this notation $NNx = 3\text{–}Nx = 3\text{–}3 + x = x$; $A1N1 = A1(3\text{–}1) = min(1, 2) = 1$, $A2N2 = A2(3\text{–}2) = min(2, 1) = 1$, etc. Moreover, it is always possible to establish a correspondence between the different notations, so that the laws will appear as invariant with respect to these notations, i.e. as not depending on their differences.

§ 3. Two-valued logic includes those logical systems which are based on the hypothesis of the two-valuedness of propositions. We can here speak of systems (and not only of one system), because in constructing a two-valued algebra of propositions different sets of functions can be taken as basic (cf. HILBERT and ACKERMANN). In this sense the term 'two-valued logic' means 'two-valued logical system' or 'the totality of two-valued logical systems'. But that is not all. By two-valued logic we also mean the investigation of the properties and relations between two-valued logical systems. This is fully justified (and corresponds to the factual use of the term), because the task of logical investigations is not only the construction of different logical systems, one after another. Finally, in speaking of two-valued logic one can mean a conception of logic as a whole, according to which the hypothesis of the two-valuedness of propositions is explicitly or implicitly acknowledged as basic for logic in general. This two-valued conception of logic is expressed, for instance, by constructing the propositional calculus, understood as the basis of the theory of deduction in general, in such a way that all tautologies of the two-valued algebra of propositions are derivable in it, so that it is deductively complete with respect to the two-valued algebra of propositions (cf. NOVIKOV). To speak of a two-valued conception of logic seems appropriate also because there are two-valued logical systems which occur in subordinated positions (the logic of predicates includes the logic of propositions, the logic of probability presupposes an ordinary logic of propositions, etc.).

In this last sense one speaks also of classical logic. But in our opinion, classical logic (or the classical conception of logic) has further historical origins besides the hypothesis of two-valuedness (it is highly possible, that from an historical point of view the hypothesis of two-valuedness is a conclusion from these other origins). Therefore it will be more expedient to avoid the use of terms that call forth associations which, although fully justified from the point of view of the history of logic or a broad investigation of the contemporary state of logic, are nevertheless superfluous in our case. Thus, by the two-valued conception of logic we will mean the consequences of the hypothesis (or the principle, the postulate, etc.) of two-valuedness on the construction of the science of logic as a whole, at least on the construction of the theory of deduction which forms the nucleus of logic.

We have enumerated these three senses of the term 'two-valued logic' primarily because we wish to define the term 'many-valued logic' in a similar way. By many-valued logic we will understand first of all the totality of logical systems (theories, constructions), whose propositions may be assigned more than two truth values, or in the general case, any finite or even infinite number of truth values.

The traditional 'true' and 'false' appear only as particular cases of such values. By many-valued logical systems we mean many-valued constructions in the logic of propositions and predicates. As a branch of scientific investigation many-valued logic is, of course, not only the totality of logical systems based on the hypothesis of the many-valuedness of propositions. It includes also the general problems of constructing such logical systems, the investigation of their properties and relations, the discovery of general methods for solving their distinctive problems (e.g. the problem of axiomatization), etc. – in short, it includes those theoretical investigations whose subject are the many-valued constructions themselves. Finally, one can speak of many-valued logic as a conception of logic as a whole (namely as a conception of logic which admits the possibility not only of a two-valued but also of a three or more valued approach to propositions). We might note that from the fact that a given conception of logic is not a two-valued one it does not follow that it must be a many-valued one: there are constructions of logic which are not based on the possible truth values of propositions and for many parts of logic it is not necessary to check the truth values of the component propositions (cf., e.g., ZINOV'EV 1961a).

These definitions of two-valued and many-valued logic are not to be estimated on the same grounds as the definitions included in logical theories. They only indicate somehow the direction of the further considerations without influencing their character.

The main point in the definition of many-valued logic mentioned above is the reference to the number of truth values of propositions (or to the number of subsets into which the set of propositions is divided). If this point is not kept in mind, then the boundary between two-valued and many-valued logic becomes wholly indeterminate. We shall see later that an axiomatization of the classical propositional calculus can be interpreted by many-valued matrices, that many functions of many-valued systems can similarly be defined in two-valued ones, that axiomatic constructions in many-valued logic satisfy two-valued matrices, etc.

Just as the term 'classical' is associated with two-valued logic, similarly the term 'non-classical' is associated with many-valued logic. But in the latter case, as in the former, the two notions are not identical: from a historical point of view the appearance of 'non-classical' logical systems (e.g. the constructivist propositional calculus, the logic of Lewis, etc.) is due to causes other than the admission of more than two truth values. There are 'non-classical' logical systems which do not presuppose any hypothesis about the number of truth values (cf. ZINOV'EV 1961a). Therefore, in the following we will not use the term 'non-classical'.

In defining many-valued logic we intentionally spoke only about the possibility of assigning more than two truth values to the propositions, thereby not excluding the two-valued approach itself as a particular case. The fact is that the available many-valued systems can be divided into three groups according to their basic hypothesis concerning the number of truth values (the division is simply given by enumeration and does not satisfy strict rules of division, these being unnecessary here). To the first group belong the systems in which the number of truth values is an exactly fixed integer larger than two, namely the three-, four-, six-, etc., valued systems. One often means by many-valued logic logical constructions of just this type, stressing the presence of additional truth values besides 'true' and 'false'. To the second group belong the systems in which the number of truth values is defined by a class of numbers, e.g. by 2^n, $3n$, n^2, etc. Obviously, any such construction represents in fact a class of systems of the first group. Another example is the class of logical systems obtained by multiplying matrices according to the method of Jaśkowski, which we will consider below. The construction of the systems of this group follows a more general approach than was the case for the construction of the systems of the first group. But the approach is not yet fully general. A completely general approach is found in the construction of the systems of the third group where an arbitrary finite or infinite set of truth values is assumed. It is obvious that the systems of the third and part of the systems of the second group include two-valued logic as a particular case; this follows from the very hypotheses on the number of truth values. The sense in which other systems generalize two-valued logic will be explained below.

The definitions of two-valued and many-valued logic could be formulated in a more general terminology. The first would speak of assuming two and only two mutually exclusive possibilities in general, and the second, of any number of such possibilities. For instance, we could start by assuming a number of possible states of an object ('an object either has some property or does not have it', 'an object can be in one of three given states', etc.), and make use of a corresponding ontological terminology. But, as has been mentioned already, this question is not one of principle: it is always possible to translate such constructions into a language using the terms 'proposition', 'truth value', etc.

§ 4. As remarked rightly by Rosser and Turquette (p. 10): 'Ever since there was first a clear enunciation of the principle "Every proposition is either true or false", there have been those who questioned it.' These doubts have had a reasonable and real sense. The principle just mentioned led to difficulties with respect to the estimation of the truth values of propositions concerning future events, of propositions in which the time or place of the events is not indicated, of propositions obtained from mutually exclusive experiences, etc. Similar difficulties arose in the attempt to construct a modal logic. These facts are pointed out in ROSSER and TURQUETTE, KOTARBIŃSKI, GRENIEWSKI 1955, MOSTOWSKI and many other works. But as long as they were not embodied in the form of complete logical systems, doubts of this kind had only the significance of historical facts, but nothing more. Certain conditions had to be realized first within logic itself, before these doubts could play the rôle of stimuli in the construction of many-valued systems.

By internal conditions within logic itself, we mean here the elaboration of the contemporary methods of logical investigation (above all: the matrix method and the ability for a purely formal approach to logical problems) and the accumulation of experience in the construction and investigation of logical systems within the limits of two-valued logic.

It was not by chance, therefore, that many-valued logic originated and developed comparatively late, namely in the twenties of our century. Its founders were Łukasiewicz (1920)[7] and Post (1921). An important rôle in the development of many-valued logic has been played by the ideas of Brouwer (1924), which are sometimes referred to as a third source of many-valued logic (cf. KOTARBIŃSKI, ZINOV'EV 1960).

§ 5. Historically the first many-valued system (propositional logic) is the system constructed by Łukasiewicz (cf. ŁUKASIEWICZ 1920, 1920a, 1930, ŁUKASIEWICZ and TARSKI). Starting with the analysis of modal propositions, Łukasiewicz came to the conclusion that two-valued logic is insufficient for the description of the mutual relations and properties of these propositions, that we need here a logic in which, besides the classical truth values 'true' and 'false', there is a third value 'possible', 'neutral' (a neutral, intermediary value). It should be stressed that 'possible' here is not a modal functor entering the internal structure of a proposition. But it is an evaluation of a proposition in its relation to reality, lying outside the proposition itself and not entering its internal structure, in the same way in which the evaluation of propositions by the terms 'true' and 'false' does not enter the internal structure of the propositions in question. Of course, the terms 'true', 'false' and 'possible' can occur as predicates in propositions of the type 'proposition x is true (false, possible)', but this does not change the main point: if x is some proposition, then its evaluation by the terms mentioned leads to the formation of a new proposition in which the name of proposition x appears only as the subject.

Treating many-valuedness as a division of the set of propositions not into two but into three and more non-overlapping subsets (the division is assumed to be exhaustive), we have in the Łukasiewicz system three classes of propositions (GRENIEWSKI 1957, SUSZKO):

1) true ones,
2) false ones,
3) neutral ones,

so that for every proposition the following principle is valid: 'the proposition is either true or false or neutral'.

The conclusion that two-valued logic is insufficient for the description of modal propositions seems to be supported by the following fact. In two-valued logic the conjunction of the propositions 'it is possible that *x*' and 'it is possible that not-*x*' (here 'possible' is a modal functor, a symbol of a modality) should, on the one hand, be taken as false, namely if the propositions are considered in analogy to affirmation and negation, whereas on the other hand, from an informal point of view, the possibility of its truth does not seem doubtful: there are events, to which *x* may refer, which possibly may occur or not occur.

Let us adopt the following symbolism:

Mx – it is possible that *x*,
NMx – it is impossible that *x*,
MNx – it is possible that not-*x*,
NMNx – it is impossible that not-*x* (it is necessary that *x*), where after the word 'that' the content of *x* is expounded.

We assume further that the modal propositions include the propositions characterized by the following assertions:

1) *CNMxNx* – if it is impossible that *x*, then not-*x*,
2) *CNxNMx* – if not-*x*, then it is impossible that *x*,
3) Σ*xKMxMNx* – there is an *x* for which it is possible that *x* and possible that not-*x*.

Within the limits of two-valued logic, the admission of these assertions leads to contradiction. In particular, the assertions *CMxx* and *CMNxNx* (if it is possible that *x*, then *x;* if it is possible that not-*x*, then not-*x*) become valid there; but this means that in the case where *Mx* and *MNx* are both true, *x* and *Nx* would both have to be true, which is contradictory; the following deductions by *modus ponens* give

$$\frac{\begin{array}{c} CMxx \\ Mx \end{array}}{x} \qquad \frac{\begin{array}{c} CMNxNx \\ MNx \end{array}}{Nx}$$

but, according to the third assertion, the conjunction of these conclusions is valid for some x.

For the construction by matrices, the situation is similar. In two-valued logic four functions of one argument x are possible:

x	U^1x	U^2x	U^3x	U^4x
1	0	1	0	1
0	0	0	1	1

Mx should be identical with one of them. But assertions (1)–(3) exclude this. The first is valid (true) only in the case where $Mx = x$ or $Mx = U^4x$; the second – in the case where $Mx = x$ or $Mx = U^1x$; the third – in the case where $Mx = U^4x$. That the third assertion is valid only in the case where $Mx = U^4x$, can be verified under the assumption that $\Pi x\alpha(x) = K\alpha(0)\alpha(1)$, where Π is the universal quantifier and α any function; because, as $\Sigma xKMxMNx = N\Pi xNKMxMNx$, the third assertion becomes then equivalent with $KM0M1$, which is possible only in the case where $M0 = M1 = 1$.

Thus an examination shows that the first and the second assertion are true if $Mx = x$; the first and the third – if $Mx = U^4x$; whereas the second and the third are incompatible. Therefore, there is no function for Mx which would satisfy assertions (1)–(3).

It should be stressed that, in speaking of the insufficiency of two-valued logic, we mean here only an insufficiency from a determinate point of view (namely from the point of view of the assumptions concerning modal propositions mentioned above) and not a general insufficiency.

In ŁUKASIEWICZ 1951, Łukasiewicz revised his point of view on modal logic and applied instead of a three-valued logic a four-valued one in which the laws of two-valued propositional logic remain valid. But this circumstance does not affect what we said above, because here we are interested only in illustrating a general idea and not in deciding on the merits and shortcomings of different approaches to modal logic. Moreover, the earlier works of Łukasiewicz will never lose their significance as a source of the very idea of many-valued logic.

The four-valued logic which Łukasiewicz used in ŁUKASIEWICZ 1951 is constructed by multiplying the two-valued matrices of implication and negation with themselves. (This method of multiplying matrices will be treated in § 3 of the following chapter).

It is not necessary here to describe the mutual relations of modal propositions treated in the work of Łukasiewicz, since it is of historical interest only. The essential thing is the proof itself that a non-two-valued logical system is possible. The great significance of this latter achievement can hardly be overestimated.

The system of Łukasiewicz is constructed as follows: The truth values are designated by the symbols *1* (true), *0* (false) and $\frac{1}{2}$ (the third value). The negation Nx of the proposition x and the implication Cxy of the propositions x and y are defined according to the matrices

x	Nx
1	0
0	1
$\frac{1}{2}$	$\frac{1}{2}$

x \ y	1	0	$\frac{1}{2}$
1	1	0	$\frac{1}{2}$
0	1	1	1
$\frac{1}{2}$	1	$\frac{1}{2}$	1

The disjunction Axy and the conjunction Kxy can be defined with the help of implication: $Axy = CCxyy$, $Kxy = NCCNxNyNy$ or $Kxy = NANxNy$ (here the symbols N, A, K, and C denote, of course, three-valued functions; making suitable definitions and conventions, we will use these symbols in logics with any number of truth values). But disjunction and conjunction can also be defined by corresponding matrices:

x \ y	1	0	$\frac{1}{2}$
1	1	1	1
0	1	0	$\frac{1}{2}$
$\frac{1}{2}$	1	$\frac{1}{2}$	$\frac{1}{2}$

x \ y	1	0	$\frac{1}{2}$
1	1	0	$\frac{1}{2}$
0	0	0	0
$\frac{1}{2}$	$\frac{1}{2}$	0	$\frac{1}{2}$

In this case the defining equations given above will become assertions concerning the equivalence of propositions which are verified by the matrices.

The functions under consideration are written in the form of equations as follows:

1) $Nx = 1 - x$;
2) $Cxy = 1$, if $x \leqslant y$; $Cxy = 1 - x + y$, if $x > y$; $Cxy = min(1, 1 - x + y)$;
3) $Axy = max(x, y)$;
4) $Kxy = min(x, y)$.

One more advantage of this notation becames apparent here: by means of it one can easily make generalizations to any finite or denumerably infinite set of values similar to those made by Łukasiewicz. For instance, taking as truth values the rational numbers from *0* to *1*, we get the following relations: if $x = \frac{3}{7}$, then $Nx = 1 - \frac{3}{7} = \frac{4}{7}$; if $x = \frac{3}{7}$ and $y = \frac{2}{7}$, then $Cxy = min(1, 1 - \frac{3}{7} + \frac{2}{7}) = min(1, \frac{6}{7}) = \frac{6}{7}$.

Since conjunction and disjunction are defined with the help of implication and negation, the system of Łukasiewicz (more exactly: of Łukasiewicz and Tarski, because its axiomatization is due to Tarski) is known as the system *N–C*, i.e. as the system based on the operators *N* and *C* (on the primitive functions *Nx* and *Cxy*). If one chooses as truth values the positive integers from *1* to *n*, then the defining equations of *N* and *C* will have a somewhat different form, but this will not change the main point.

As in two-valued logic, the tautologies (the asserted, proved, always-true, etc., propositions) of the system of Łukasiewicz are those propositions which have value *1* under all possible substitutions, i.e., for any values whatsoever of the arguments (of the primitive propositions out of which the given proposition is formed).

With the help of the matrices (or of calculation by means of substituting truth values in equations equal to the matrices) one can verify that not all tautologies of two-valued logic are tautologies in the logic of Łukasiewicz. Thus *CCNxxx*, *NKxNx*, *AxNx* are tautologies (laws) of two-valued logic, but are not tautologies in the logic of Łukasiewicz, since respectively

$CCN\frac{1}{2}\,\frac{1}{2}\,\frac{1}{2} = CC\frac{1}{2}\,\frac{1}{2}\,\frac{1}{2} = C1\,\frac{1}{2} = \frac{1}{2}$;
$NK\frac{1}{2}N\frac{1}{2} = NK\frac{1}{2}\,\frac{1}{2} = N\frac{1}{2} = \frac{1}{2}$; $A\frac{1}{2}N\frac{1}{2} = A\frac{1}{2}\,\frac{1}{2} = \frac{1}{2}$.

Also, the principles of *reductio ad absurdum* of two-valued logic, *CCxNxNx* and *CCxKyNyNx*, are here not tautologies, since for $x = y = \frac{1}{2}$ they take the truth value $\frac{1}{2}$. *CNxy* and *CCxyy*, which are equivalent in two-valued logic, are not equivalent here, because $CN\frac{1}{2}\,\frac{1}{2} = C\frac{1}{2}\,\frac{1}{2} = 1$ and $CC\frac{1}{2}\,\frac{1}{2}\,\frac{1}{2} = C1\,\frac{1}{2} = \frac{1}{2}$. The case is similar with *Cxy* and *ANxy*, since $C\frac{1}{2}\,\frac{1}{2} = 1$ and $AN\frac{1}{2}\,\frac{1}{2} = A\frac{1}{2}\,\frac{1}{2} = \frac{1}{2}$. On the other hand, for example, *CNNxx* and *CxNNx* prove to be tautologies here as well, since by definition $NNx = 1 - Nx = 1 - 1 + x = x$, i.e. the double negation of any truth value results in the same truth value (this in distinction to intuitionist logic, where, as we will see below, only the second of the two formulas is a law).

It is essential to note the following property of the system of Łukasiewicz. The propositions *NKxNx* and *AxNx* correspond to the law of contradiction and the law of the excluded middle of two-valued logic. As we saw above, they are not laws in the system of Łukasiewicz. This means nothing more than the following: if $x = \frac{1}{2}$, then the given propositions as a whole also take the value $\frac{1}{2}$, i.e., they do not always take the value *1*.

On the other hand, the negations of the law of contradiction and of the law of the excluded middle are also not laws in the system of Łukasiewicz, since $NNK1N1 = 0$, $NNK0N0 = 0$, $NNK\frac{1}{2}N\frac{1}{2} = \frac{1}{2}$; $NA1N1 = 0$, $NA0N0 = 0$, $NA\frac{1}{2}N\frac{1}{2} = \frac{1}{2}$. It is also very important to note the following fact: the logic of Łukasiewicz is not the negation of two-valued logic, but a generalization of it. Indeed, if we consider only *0* and *1*, then we can read out of Łukasiewicz's matrices the matrices of ordinary two-valued negation, implication and other functions. In other words, if $x = 1$ or $x = 0$ and all intermediary values are excluded, then two-valued logic appears as a particular (limiting) case of the system of Łukasiewicz. In particular $N1 = 1 - 1 = 0$, $N0 = 1 - 0 = 1$.

The system of Łukasiewicz has been axiomatized by Tarski and by Wajsberg (cf. Łukasiewicz and Tarski, Wajsberg). We will examine in more detail below the relation between matric (truth functional) and axiomatic constructions. Here we limit ourselves to short remarks. The axioms of Wajsberg are the following:

1) *CxCyx*,
2) *CCxyCCyzCxz*,
3) *CCCxNxxx*,
4) *CCNyNxCxy*.

Axioms (1), (2) and (4) are tautologies of two-valued logic. In place of axiom (3), two-valued logic has *CCNxxx*, which is not a tautology in the system of Łukasiewicz, since $CCN\frac{1}{2}\,\frac{1}{2}\,\frac{1}{2} = \frac{1}{2}$. By means of the matrices (and the equations formulated above) one can verify that axioms (1)–(4) are all tautologies, i.e., they always take the value *1*. Generally, in choosing an axiom system one takes propositions asserted in the matrix construction, in the case given here – always true propositions. In the same way one can also verify that the given axiom system does not contradict two-valued logic: the third axiom takes the value *1*, if *x* has either one of the values *1* or *0*, since $CCC1N111 = CCC1011 = CC011 = C11 = 1$, $CCC0N000 = CCC0100 = CC100 = C00 = 1$. The third axiom is a tautology in two-valued logic. Thus, an axiom system intended for a many-valued logic can in many cases be interpreted as two-valued.

Łukasiewicz linked the idea of many-valued logic with the calculus of probability: if the third value ('possible') has gradations, then an infinite number of values (similar to probability within the limits from *0* to *1*) is possible.

Three-valued and infinite-valued logics are, for Łukasiewicz, fragments of two-valued logic in the sense that the assertions of many-valued logic (without quantifiers) are valid also in two-valued logic (the converse, as we have seen, does not always hold). In three-valued logic there are assertions not valid in infinite-valued logic.

Łukasiewicz remarks justly, that it would be inaccurate to call many-valued logic non-Aristotelian: Aristotle himself admitted many propositions (propositions about future events) to be neither true nor false. It would be more correct to call it non-Chrysippean, since Chrysippus was the first to declare categorically that all propositions are either true or false (cf. ŁUKASIEWICZ 1930).

§ 6. Post published his many-valued system shortly after Łukasiewicz. He began, in distinction from Łukasiewicz, with purely formal considerations: he allowed the arguments to take values out of a given number of *n* values (say *1, 2, ..., n*), and considered functions of such arguments as taking their values out of the same set of *n* values. The question of what meaning the expression 'value *i*' could thereby have, did not interest Post at all. Clearly, such an approach was only possible because it had already become familiar in connection with two-valued logic: there the ability of abstracting from what the expressions 'true' and 'false' signify had been developed, as well as replacing these words by any suitable symbols. In this way attention could be concentrated on the purely logical relations and instead of propositions one could take any arguments and functions defined by these logical relations. The approach chosen by Post precluded the consideration of all philosophical problems.

But among other things it should be remarked that although Łukasiewicz did start from some informal problem, the process itself of constructing a many-valued system at that point demanded a formal approach (i.e. an abstraction from the meaning of the symbols designating the truth values). In Post this differentiation between the process of constructing the logical system and the attribution of an informal interpretation is absolutely sharp: he gives no interpretation at all. Such a type of abstraction is, however, not only permissible, but even necessary for mastering the technical (if it can be thus expressed) difficulties of constructing many-valued systems.

Post constructs his many-valued system as a generalization of a two-valued one. This must not be interpreted, as if all functions of many-valued logic would have analogs in two-valued logic (we will, see below that there are functions for which there are no such analogs). It must be understood in such a way that for *n* equal to two, we will get two-valued logic as a particular case.

Post defines the negations by the matrices:

x	N^1x
1	2
2	3
⋮	⋮
n–1	n
n	1

x	N^2x
1	n
2	n–1
⋮	⋮
n–1	2
n	1

or, in another form, by the equations

1) $N^1x = x + 1, \qquad N^1n = 1,$

2) $N^2x = n - x + 1.$

He defines disjunction in the same way as Łukasiewicz, i.e., by $Axy = min(x, y)$, with only the difference that it is the minimal value which is taken as the more asserted. This is connected with the choice of the system of notation for the truth values and has no principal meaning. It is important only that, if *min* is chosen for one of the functions Axy and Kxy, then one must take *max* for the other; and that, in analogy with two-valued logic, the more asserted (true) value is chosen correspondingly.

Conjunction is defined by disjunction and negation: $Kxy = N^2AN^2xN^2y$. It can be proved that $Kxy = max(x, y)$. In fact, $Kxy = N^2AN^2xN^2y = n - AN^2xN^2y + 1 = n - min(N^2x, N^2y) + 1 =$

$= n\text{-}min(n\text{-}x + 1, n\text{-}y + 1) + 1$; suppose $x = y$, then $Kxy =$ $= n\text{-}n + x\text{-}1 + 1 = n\text{-}n + y\text{-}1 + 1 = x = y$; suppose $x>y$, then $Kxy =$ $= n\text{-}n + x\text{-}1 + 1 = x$; suppose $x<y$, then $Kxy = n\text{-}n + y\text{-}1 + 1 = y$.

Thus, we have here a generalization of the laws of two-valued logic concerning the mutual relations of conjunction and disjunction.

From the examples given it is also obvious that the generalization of the functions of two-valued logic can take different lines. Thus, both negations defined above are generalizations of two-valued negation. Yet the first must be applied *n* times in order to get the original proposition, whereas for the second negation the principle of double negation remains valid for any *i*. In the case where *n* equals two, these two negations coincide.

The rôle of the formal approach to the elaboration of logical problems hardly needs to be demonstrated to anybody nowadays. Indeed, certain real historical motives (as in the case of Łukasiewicz and, we will see below, of Brouwer and Heyting), real needs of science, are discernible in the origins of formal constructions. But the other side of the history of science is not less interesting, where the presence of some theory creates the demand for it; and how consequently different interpretations of the formal constructions originate. In these kinds of cases an approach to the solution of logical problems like that chosen by Post plays an important rôle and is invaluable as a preliminary condition for the solution of scientific problems.

At present, the situation is still such that the demand for many-valued logic is relatively small in comparison with all that has been done in this field from a purely formal point of view. Pioneer work is done especially in the line of elaborating logic as a formal apparatus. From this point of view the work of Post plays an important rôle.

CHAPTER TWO

OUTLINE OF MANY-VALUED LOGICAL SYSTEMS

§ 1. It is impossible to somehow give here a complete survey of all the works on many-valued logic that have appeared after the works of Łukasiewicz and Post. And there is no necessity for this. From the point of view of our subject it will be sufficient in what follows to consider a selection of various other systems, with the intention of using them to illustrate further aspects of this subject and not worrying about observing the rules of historical investigation. The development of many-valued logic unfolded and still evolves mainly in the following three directions:

1) purely formal elaboration of the logical apparatus;

2) construction of logical systems, or adaption of systems already available, for the purpose of solving concrete problems of scientific investigation;

3) elaboration of a general theory of many-valued logical systems.

However these are aspects of a single process and not directions isolated from one another. In many works, different sides of the matter are touched upon either directly or indirectly, so that it is hardly possible to establish any strict classification.

§ 2. As was mentioned already, an important rôle in the development of many-valued logic has been played by the ideas of Brouwer (cf. BROUWER 1923, 1925; see also HEYTING 1930, 1956, 1956a, KOTARBIŃSKI, ZAWIRSKI). The following statement of Brouwer served as a starting-point: the general validity of the law of the excluded middle is limited to that part of mathematics (and this means also to that part of natural science) which is developed within a determinate, finite mathematical system (i.e., onto which a determinate, finite mathematical system can be projected).

In ZINOV'EV 1960 we treated the works of Brouwer as a third primary source of the idea of many-valued logic in contemporary logic. However, if one considers only those logical systems as strictly many-valued, in which the corresponding hypothesis concerning the number of truth values of propositions 'operates' (i.e., those constructed by matrices), then our assertion was false. Without undue emphasis, however, we should note the following. The law of the excluded middle, formulated in the form 'either *x* or not-*x*' clearly expresses a hypothesis about two possibilities. And to doubt its absolute generality is to doubt the absoluteness of the hypothesis of two-valuedness. On the other hand, it was only when many-valued logic had already become a fact, that this doubt received a precise expression in logic. In this sense it is sufficient to speak only of an influence of the ideas of intuitionism on the development of many-valued logic. It is important to stress that in this case the very concrete needs of science acted as stimuli.

In two-valued logic *KCNNxxCxNNx* holds, i.e. the double negation of *x* is equivalent to *x* (in other words *CNNxx* and *CxNNx* are laws). From a logical point of view the idea of Brouwer amounts in fact to the following: to 'weaken' this law by excluding one part of it, *CNNxx*, while retaining the other part, *CxNNx*, (using his terminology: to reject the principle according to which *x* follows from the absurdity of the absurdity of *x*). Heyting (cf. HEYTING 1930, 1956, 1956a) has formulated an axiom system for the logic of propositions on this basis:

1) *CxKxx*,
2) *CKxyKyx*,
3) *CCxyCKxzKyz*,
4) *CKCxyCyzCxz*,
5) *CyCxy*,
6) *CKxCxyy*,
7) *CxAxy*,
8) *CAxyAyx*,
9) *CKCxzCyzCAxyz*,
10) *CNxCxy*,
11) *CKCxyCxNyNx*.

All theorems of this system are tautologies if they are evaluated according to the following three-valued matrices:

x	Nx
1	0
0	1
$\frac{1}{2}$	0

x \ y	1	0	$\frac{1}{2}$
1	1	0	$\frac{1}{2}$
0	1	1	1
$\frac{1}{2}$	1	0	1

The tables of conjunction and disjunction are the same as in Łukasiewicz. In equation form, implication can be given by $Cxy = 1$, if $x \leqslant y$, and $Cxy = y$, if $x > y$. Thus, if in Łukasiewicz $C\frac{1}{2}\,0 = 1 - \frac{1}{2} + 0 = \frac{1}{2}$ and $C1\,\frac{1}{2} = 1 - 1 + \frac{1}{2} = \frac{1}{2}$, then according to these Heyting matrices, as we will call them, $C\frac{1}{2}\,0 = 0$ and $C1\,\frac{1}{2} = \frac{1}{2}$. We thus see that many-valued systems can be constructed in different ways.

According to the given Heyting matrices *CNNxx* and *AxNx*, for example, are not tautologies: $CNN\frac{1}{2}\ \frac{1}{2} = CN0\ \frac{1}{2} = C1\ \frac{1}{2} = \frac{1}{2}$ and $A\frac{1}{2}N\frac{1}{2} = A\frac{1}{2}\,0 = \frac{1}{2}$. But *CxNNx* comes out to be a tautology, as $C1NN1 = 1$, $C0NN0 = 1$, $C\frac{1}{2}NN\frac{1}{2} = C\frac{1}{2}N0 = C\frac{1}{2}\,1 = 1$; similarly all axioms of the Heyting system; e.g., axiom (7): if $x \geqslant y$, then by definition $Axy = x$, and $Cxx = 1$; if $x < y$, then $Axy = y$, and $Cxy = 1$.

But although all theorems of the Heyting system become tautologies according to these matrices, the matrix construction is not equivalent with the axiom system (cf. ZAWIRSKI, KLEENE, KOTARBIŃSKI). There are formulas which are tautologies according to the matrices and which are not theorems of the Heyting system: e.g., *ACxyCyx*, *CCNNxxAxNx*.[8]) Gödel maintains (cf. GÖDEL) that the intuitionistic calculus of propositions is such that no truth table construction using a finite number of truth values will ever adequately characterize it.

Another axiom system of intuitionistic propositional calculus equivalent to the Heyting system has been formulated by Łukasiewicz:

1) *CxCyx*,
2) *CCxCxyCxy*,
3) *CCxyCCyzCxz*,
4) *CKxyx*,
5) *CKxyy*,
6) *CCxyCCyzCxKyz*,
7) *CxAxy*,
8) *CyAxy*,
9) *CCxzCCyzCAxyz*,
10) *CCxNyCyNx*,
11) *CNxCxy*.

Here, it is sufficient to add the axiom

12) *CCCxNxyCCxyy*

in order to get the two-valued calculus of propositions, i.e., to get a construction in which all the tautologies and only the tautologies of the two-valued algebra of propositions are derivable (normally, *modus ponens* and a rule of substitution are accepted as rules of derivation, and therefore we will not mention them every time).

A further axiomatization equivalent with the Heyting system is the one formulated by Gentzen (cf. GENTZEN, KLEENE). If there the axiom *CNxCxy* is replaced by the axiom *CNNxx* (i.e., if one accepts that part of the law of the equivalence of double negation and affirmation which Brouwer rejected), then we get an axiom system of the two-valued calculus of propositions. We have here a replacement of an axiom instead of an addition, but the result can be shown to be the same as before.

It would be a mistake to think that in the Brouwer-Heyting system the law of the excluded middle is false and its negation true. More than that, it would be a great mistake to think that the doubts concerning its general validity signify that some instances of the negation of this law would be accepted. In the logic of Brouwer-Heyting, all instances of *NANxx* are false; also, according to the Heyting matrices this formula is a contradiction. (In the logic of Łukasiewicz it can have the value $\frac{1}{2}$, but if values corresponding to truth and falsity are substituted, then it is false). Neither the logic of Heyting, nor the many-valued construction based on the Heyting matrices, contradict ordinary two-valued logic. In the second there even occurs a tautology which can be considered as a peculiar generalization of the law of the excluded middle, namely *AAxNxNNx* (the formula is not a law of the logic of Heyting): $AA1N1NN1 = AA101 = A11 = 1$, $AA0N0NN0 = AA010 = A10 = 1$, $AA\frac{1}{2}N\frac{1}{2}NN\frac{1}{2} = AA\frac{1}{2}\,01 = A\frac{1}{2}\,1 = 1$. This formula can be read as '*x* or not-*x* or not-not-*x*'; since double negation is not equivalent with assertion, the member 'not-not-*x*' cannot be dropped from the disjunction. In the following we will still add several new aspects to what has already been said about the generalization of two-valued logic, in order to show that it cannot be reduced to any single type of relation.

Generally speaking, the question now arises: is a logical system possible in which the negations of the law of the excluded middle and of the law of contradiction of two-valued logic are laws? We will come back to this question below. But on the basis of what has been said, the following can be remarked: as far as *n*-valued logical systems are generalizations of the two-valued one, then for *n* equal to two, the laws considered should have a value corresponding to truth and their negations a value corresponding to falsity.

Here it must be mentioned that, independently of Heyting, Kolmogorov (cf. KOLMOGOROV 1925, 1932) sketched a formal axiomatic apparatus for logic which does not utilize the law of the excluded middle, and that Glivenko (cf. GLIVENKO 1928, 1929) developed the idea of Kolmogorov, calling the axiom system for the propositional calculus which he obtained, 'constructivist'.

In the subsequent developments, the axiom system of the constructivist propositional calculus took the following form:

1) *CxCyx*,	6) *CCxyCCzyCAxzy*,
2) *CCxCyzCCxyCxz*,	7) *CxAxy*,
3) *CxCyKxy*,	8) *CyAxy*,
4) *CKxyx*,	9) *CCxyCCxNyNx*,
5) *CKxyy*,	10) *CNxCxy*.

The derivation scheme has the form:

$$\frac{\begin{array}{c} Cxy \\ x \end{array}}{y}$$

The constructivist and the Heyting axiom system are equivalent.

§ 3. In the current of ideas of intuitionistic logic, the method proposed by Jaśkowski (cf. ZAWIRSKI) for constructing matrices for $n + 1$ values on the basis of matrices for n values and also the method of multiplying matrices presents some interest. Of course, in both cases for $n>2$ the process must include two-valued logic as one case, so that consideration of these methods reveals one more aspect of the mutual relation between two-valued and many-valued logics.

The following correlations are given:

a)	b)	c)
1	1	1
$\alpha(1) = 0$	$\alpha(1) = 2$	$\alpha(1) = 3$
	$\alpha(0) = 0$	$\alpha(2) = 2$
		$\alpha(0) = 0$

etc., where $0, 1, 2,\ldots$ are truth values, and $\alpha(0), \alpha(1), \alpha(2),\ldots$ their negations. The lists (a), (b), (c),... correspond respectively to two-valued, three-valued, four-valued, etc. logic. If we designate negation, implication, conjunction and disjunction in n-valued logic respectively by N, C, K and A, and in $(n + 1)$-valued logic respectively by N^*, C^*, K^* and A^*, then the schemata for defining the second by means of the first appear as follows:

a)

N*	
1	α(N1)
α(x)	Nx

b)

C*	1	α(y)
1	C11	α(C1y)
α(x)	Cx1	Cxy

c)

K*	1	α(y)
1	K11	α(K1y)
α(x)	α(Kx1)	α(Kxy)

d)

A*	1	α(y)
1	A11	A1y
α(x)	Ax1	α(Axy)

where x is the first and y the second argument (the order is important for implication).

We illustrate the functioning of the schemata by the example of negation and implication. For one-valued logic, where $x = Nx = N1 = 1$, we get, of course,

N*	
1	α(1)
α(1)	1

In two-valued logic, where $\alpha(1) = 0$, we get the matrix

N*	
1	α(1) = 0
α(1) = 0	N1 = 1

In three-valued logic, where $x = 0$ and $x = 1$ have to be considered and where $\alpha(1) = 2$, $\alpha(0) = 0$, we get the matrix

N*	
1	α(0) = 0
α(0) = 0	N0 = 1
α(1) = 2	N1 = 0

For implication of one-valued logic we get the matrix

C*	1	α(1)
1	C11	α(C11) = C11
α(1)	C11	C11

or

C*	1	α(1)
1	1	α(1)
α(1)	1	1

In two-valued logic, where $\alpha(1) = 0$, we get the matrix

C*	1	α(1) = 0
1	C11 = 1	α(C11) = 0
α(1) = 0	C11 = 1	C11 = 1

In three-valued logic, where $x = 1$, $x = 0$, $y = 1$ and $y = 0$ have to be considered, and where $\alpha(1) = 2$ and $\alpha(0) = 0$, we get the matrix

C*	1	α(1) = 2	α(0) = 0
1	C11	α(C11)	α(C10)
α(1) = 2	C11	C11	C10
α(0) = 0	C01	C01	C00

i.e.

C*	1	2	0
1	1	2	0
2	1	1	0
0	1	1	1

It is interesting to note with reference to these matrices that by replacing the highest value of an n-valued[9]) matrix by the value 1 we get the corresponding $(n-1)$-valued matrix (as can easily be verified, e.g. in the case of the matrix last given). A similar replacement of the value $\frac{1}{2}$ by the value 1 in the matrices of Łukasiewicz would lead to absurdity, since we would get $C10 = 1 = 0$. If a matrix for $n + 1$ values is derived from a matrix for n values according to the method of Jaśkowski, then the set of logical expressions satisfying the derived matrix will also satisfy the other matrix. Thus, in one-valued logic the principles of two-valued logic are valid, in two-valued logic those of three-valued logic, etc.

Matrices are multiplied in the following way. Let 1 and 0 be truth values. Then the ordered pairs $(1,1)$, $(1,0)$, $(0,1)$ and $(0,0)$ can be formed. By definition we put $C(a, b)\ (c, d) = (Cac, Cbd)$ and $N(a, b) = (Na, Nb)$, where $a, b, c, d = 1,0$. Thus explaining for every case what pairs of values correspond to $C(a, b)\ (c, d)$ and $N(a, b)$, and replacing the pairs $(1, 1)$, $(1, 0)$, $(0, 1)$ and $(0, 0)$ by the symbols $1, 2, 3$ and 0 respectively, we obtain the definition of a four-valued C and N. By a similar method, eight-valued, sixteen-valued, etc., C and N can be defined, taking sequences of three values, four values, etc.

The definitions of A and K are similar, by means of the equations $A(a, b)\ (c, d) = (Aac, Abd)$ and $K(a, b)\ (c, d) = (Kac, Kbd)$. It can easily be verified, that the laws of two-valued propositional logic remain valid. If a law is a proposition always having the value 1 (true), then the laws of the excluded middle $AxNx$ and of contradiction $NKxNx$, and also the equations $Cxy = ANxy$ and $Cxy = NKxNy$ remain valid. In general, the many-valued logical constructions obtained by this method can be considered as interpretations of the principles of the two-valued logic of propositions. This shows that a many-valued logic need not necessarily reject laws of two-valued logic.

The matrix obtained by the multiplication of two two-valued matrices of negation has the following form:

N	
11 (= 1)	00 (= 0)
10 (= 2)	01 (= 3)
01 (= 3)	10 (= 2)
00 (= 0)	11 (= 1)

and the matrix obtained by the multiplication of two two-valued matrices of implication is the following:

C	11	10	01	00
11	11	10	01	00
10	11	11	01	01
01	11	10	11	10
00	11	11	11	11

The matrices obtained as a result of the first of the operations considered and the multiplications of matrices do not coincide. Thus, numbering the pairs in the matrix of implication mentioned above, we get

C	1	2	3	0
1	1	2	3	0
2	1	1	3	3
3	1	2	1	2
0	1	1	1	1

whereas according to the first method we get the matrix

C	1	2	3	0
1	1	2	3	0
2	1	1	1	0
3	1	2	1	0
0	1	1	1	1

Another example of a four-valued logic is given in RASIOWA; there the matrix of four-valued implication

C	1	2	3	4
1	1	2	3	4
2	2	1	4	3
3	1	2	1	2
4	2	1	2	1

is obtained by means of multiplying the two-valued matrices of implication and equivalence according to the method of Jaśkowski. (Tautologies always have the value.1)) The axioms of this system are

1) *CCxyCCyzCxz*,
2) *CyCxCxy*,
3) *CCCxyCCxyxx*;

(i.e., it is analogous to the system of Tarski-Bernays for the two-valued propositional calculus). It is characteristic that the set of its tautologies is equal to the sum of those tautologies of the two-valued implicative calculus which are also tautologies in the calculus whose implication sign has been replaced by the equivalence sign.

We see, even within the limits of ideas of Jaśkowski alone, that different aspects of the interrelation of many-valued and two-valued logic can be distinguished, as well as different ways of generalizing two-valued logic. Therefore every attempt to reduce these relations to a single form must break down as not corresponding to the facts.

§ 4. Bočvar constructed his three-valued system (cf. BOČVAR 1938, 1943) with the aim of solving the paradoxes of classical mathematical logic by means of a formal proof showing the meaninglessness of certain propositions. Bočvar distinguishes between meaningful and meaningless propositions. A proposition is meaningful, if it is true or false. Bočvar calls such a proposition '*predloženie*'.[10]) Further he distinguishes between

1) the internal form of assertion, negation, conjunction, disjunction and implication: *x*, not-*x*, *x* and *y*, *x* or *y*, if *x* then *y*, and

2) their external form: *x* is true, *x* is false, *x* is true and *y* is true, *x* is true or *y* is true, if *x* is true then *y* is true.

For meaningful propositions these two forms are equivalent. Their difference becomes apparent if one substitutes in place of *x* (or *y*) a meaningless proposition: in the first case we get a meaningless proposition, in the second case not. This difference corresponds to the difference between language and metalanguage.

For the system of Bočvar we make use of the following notation:

1) *x*, *y*, *z*,... – arbitrary propositions;
2) *1*, *2*, *3* – respectively: truth, falsity and meaninglessness;
3) *Kxy* – internal conjunction, defined by the matrix

x \ y	1	2	3
1	1	2	3
2	2	2	3
3	3	3	3

or also by the equation $Kxy = max(x, y)$;

4) *Nx*, *Vx* and *Wx* – respectively: internal negation (not-*x*), external assertion (*x* is true) and external negation (*x* is false), defined by the matrices

x	Nx	Vx	Wx
1	2	1	2
2	1	2	1
3	3	2	2

By means of these functions others can be defined:

$$
\begin{array}{ll}
 & K^*xy = KVxVy, \\
Axy = NKNxNy, & A^*xy = AVxVy, \\
Cxy = NKxNy, & C^*xy = CVxVy, \\
Rxy = KCxyCyx, & R^*xy = KC^*xyC^*yx, \\
Sxy = KRxyRNxNy, & S^*xy = KR^*xyR^*NxNy, \\
 & Tx \;\; = NAVxWx, \\
 & Ux \;\; = NVx,
\end{array}
$$

where: A, C and R are respectively symbols of internal disjunction, implication and equivalence; K^*, A^*, C^* and R^* are respectively symbols of external conjunction, disjunction, implication and equivalence; Tx signifies that x is meaningless; U is the symbol of the internal negation of external assertion.

To be a formula means

1) to be a symbol of a proposition;
2) if x is a formula, then Nx, Vx and Wx are formulas;
3) if x and y are formulas, then Kxy is a formula.

Further formulas are introduced by the definitions of new operators. A formula is proved (is a tautology) in the matrix logic of propositions of Bočvar, if it has value *1* for any values of the arguments whatsoever. A proof consists in a verification by means of the matrices. For example, the formula Rxx (the law of identity in two-valued logic), which is provable in two-valued logic, is not provable here (i.e., is not a law of Bočvar's logic), since $Rxx = KCxxCxx = KNKxNxNKxNx$, $R33 = KNK3N3NK3N3 = KN3N3 = K33 = 3$. Thus even the two-valued law of identity is not absolute. This fact plays an important rôle in Bočvar's analysis of Russell's paradox. The expression $x = x$ and the formula just mentioned are not identical.

Formulas provable in Bočvar's logic are, e.g., Sxx, $SSxNxTx$, $STxTNx$. A formula which never has the value *1*, whatever the values of its arguments may be, is called a contradiction. For example, K^*xWx, $SxWx$, R^*xUx are contradictions.

As in the case of the law of contradiction and of the law of the excluded middle (cf. § 2), the negation of the two-valued law of identity is not a law, since $NR33 = N3 = 3$, $NR11 = N1 = 2$, $NR22 = N1 = 2$. The exclusion of the two-valued law of identity from the number of laws of a given construction in no way signifies the exclusion of the principle 'x is x' or

'$x = x$'. It means only one thing: in the framework of the assumed definitions of *K*, *C*, *N*, *R*, of what is a formula and a proved formula, *Rxx* is not a proved formula. It has nothing to do with a rejection of the identity of *x* by repeated inscription or of the equation $x = x$.

In the logic of Bočvar several theorems can be derived, of which the following is important for us here: the matrix construction of the logic of propositions includes a part isomorphic with the two-valued matrix calculus of meaningful propositions. The proof is as follows. By means of the matrices we verify that the formulas given below are tautologies: *C*yC*xy*, *C*xA*xy*, *C*A*xyA*yx*, *A*xTx*, etc. (here 12 formulas are given). We establish the following correspondence: propositional variables become variables for meaningful propositions; the symbols *T*, *K**, *A**, *C** and *R** become respectively the symbols *N*, *K*, *A*, *C* and *R*. We thus get a system of formulas *CyCxy*, *CxAxy*, *CAxyAyx*, *AxNx*, etc. (12 formulas respectively) isomorphic to the system of formulas mentioned above. This system can be interpreted as an axiom system for the two-valued logic of meaningful propositions, if the following rules are accepted:

1) the rule of substitution;

2) if *x* and *Cxy* are proved formulas, then *y* is also proved;

3) if *x* and *y* are proved formulas, then *Kxy* is proved. The rule of substitution obviously remains valid. Concerning the second and the third rule, the following appears from the *C*xy* matrix: if *x* and *C*xy* are proved, then *y* is also proved; similarly for *Cxy*.

It can be shown that Bočvar's logic includes still another isomorphic image of the logic of meaningful propositions. But for us something else is here important: the relation of isomorphism is obviously a different kind of relation from the relation between the general and the particular – and this must be kept in mind in considering the mutual relation of many-valued and two-valued logic.

In Bočvar's logic the connection between two-valued and non-two-valued functions (again a new aspect of their relation) is characterized by a system of formulas. We will give some examples: *WWAxNx*, *STAxNxTx*, *WTA*xWx*; the first formula signifies that the non-two-valued negation of the law of the excluded middle is always false; the second signifies that the two-valued form of the law of the excluded middle is meaningless if and only if the proposition involved is meaningless; the third signifies that the non-two-valued form of the law of the excluded middle cannot be meaningless. The logic of Bočvar is also interesting because the consideration of many problems of the mutual relation between language and metalanguage is possible in it.

Kleene uses three-valued logic in the same way for mathematical purposes. The third value (besides truth and falsity) in KLEENE can be interpreted in different ways:

1) in the sense of 'undetermined',

2) in the sense of 'unknown whether true or false',

3) in the sense of 'the value is inessential' (we do not want to know the value, or the value plays no rôle at all),

4) in the sense of 'neither truth nor falsity can be determined algorithmically'.

To put it more exactly, one may take as truth values, for example, values of the following type:

1) 'known truth' (this corresponds to truth or, say, *1*),

2) 'known falsity' (corresponding to falsity or, say, *2*),

3) 'unknown whether true or false' (corresponding to the third value or, say, *3*),

Thus, the third value in Kleene does not appear on the same level as the others. In fact, we have the division:

1) known (determined, can be decided, etc.) which of the two values – true or false,

2) value unknown (undetermined, cannot be decided, etc.).

Since in the first case there is a division into two possibilities, all three possibilities can be considered as lying in one plane, so to speak, having associated with them the symbols *t*, *f*, *u* *(1, 2, 3)* respectively. The latter can then be considered as truth values. Thus here (as well as in Bočvar) it is not so much the case that three truth values are taken directly, but rather, we have a situation with three possibilities corresponding to which we then set up an abstract system with three values for the propositions. Of course, instead of the law of the excluded middle, a law of the excluded fourth will be valid for these possibilities. We might note that it would be more correct to formulate it here not as 'the proposition *x* has either the value *1* or *2* or *3*' but as 'concerning the proposition *x* it can be said that it either has the property *1* (namely that it is true) or the property *2* (namely that it is false) or the property *3* (its value is unknown)'.

Kleene distinguishes two types of functions (with corresponding matrices):

1) weak ones, definable by truth tables which are derived from two-valued ones by means of filling with symbols of the third value the rows and columns corresponding to this third value; e.g. for *Axy*

x \ y	1	2	3
1	1	1	3
2	1	2	3
3	3	3	3

2) strong ones, defined by the truth tables

N	
1	2
2	1
3	3

K	1	2	3
1	1	2	3
2	2	2	2
3	3	2	3

A	1	2	3
1	1	1	1
2	1	2	3
3	1	3	3

R	1	2	3
1	1	2	3
2	2	1	3
3	3	3	3

As we see, the strong tables for *N* and *R* are identical with the weak ones. Strong implication is defined by the matrix.

C	1	2	3
1	1	2	3
2	1	1	1
3	1	3	3

In the logic of Bočvar, as in the other systems, the following correlations between the truth values are established:

1) if x has the truth value *1*, then it does not have the value *2* and does not have the value *3*,

2) if x does not have the value *1*, then x has either the value *2* or the value *3* (if $n = 3$),

3) if x has the value *1* or the value *2*, then x does not have the value *3*,

4) if x does not have the value *1* and does not have the value *2*, then it has the value *3*,

and so on for the further possibilities. Taking *N*, *K* and *A* respectively as negation, conjunction and disjunction of two-valued logic, and xi as a proposition saying that x has the value i, these correlations can be written in the form: $Nx1 = x2Ax3$, $Nx2 = x1Ax3$, $x1 = Nx2KNx3$, $Nx1KNx2 = x3$, etc. In this way the logical system itself is constructed with the help of two-valued logic.

§ 5. Among the various works on many-valued logic, an important position is occupied by those works in which many-valued systems are constructed with the intention of using them to handle the philosophical and logical difficulties of quantum mechanics. These works include BIRKHOFF and VON NEUMANN, BIRKHOFF, DESTOUCHES-FÉVRIER, REICHENBACH 1944 and others. Some information on this subject can be found in KUZNECOV. We consider the system of Reichenbach to be the simplest one and it illustrates in a sufficiently clear way the ideological trend of such investigations. Concerning the general philosophical convictions of Reichenbach, we refer the reader to the great number of works, in which the philosophical errors of foreign philosophers are criticized. We limit ourselves here to critical remarks which are related to many-valued logic.

The metalanguage of the language of three-valued logic belongs to two-valued logic, i.e., propositions of the type 'x has the truth value i' are two-valued. The truth tables are constructed analogously to the two-valued ones. The defined operations are considered as generalizations of the operations of two-valued logic. The majority of operations introduced by Reichenbach were already introduced by Post, with the exception of full negation, alternative implication, quasi-implication, and alternative equivalence. The latter are introduced for purposes of quantum mechanics.

In our system of notation the functions used by Reichenbach are designated as follows:

N^1x – cyclical negation of x,
N^2x – diametrical negation of x,
N^3x – complete negation of x,
Kxy – conjunction,
Axy – disjunction,
C^1xy – standard implication,
C^2xy – alternative implication,
C^3xy – quasi-implication,
R^1xy – standard equivalence,

R^2xy – alternative equivalence,
1,2,3 – respectively truth, falsity and indeterminacy.
These functions are defined by the matrices

x	N^1x	N^2x	N^3x
1	2	3	2
2	3	2	1
3	1	1	1

x	y	Kxy	Axy	C^1xy	C^2xy	C^3xy	R^1xy	R^2xy
1	1	1	1	1	1	1	1	1
1	2	2	1	2	3	2	2	3
1	3	3	1	3	3	3	3	3
2	1	2	1	1	1	2	2	3
2	2	2	2	1	1	2	1	1
2	3	3	2	3	1	2	2	3
3	1	3	1	1	1	2	3	3
3	2	3	2	1	1	2	2	3
3	3	3	3	1	1	2	1	1

Asserted propositions are those which have the value *1* for any values of the arguments (of the primitive propositions). For saying that a proposition has a value different from *1*, the negations are used. Thus N^1N^1x signifies that x is indeterminate, N^1x that x is false, N^2x that x is false.

Among the propositions (formulas) constructed by means of the aforementioned operators, there are those which are always true, those which are always false, those which are always indeterminate, those which are true or false, those which can assume all three values, etc. It is important to note that the propositions of three-valued logic contain a subclass of propositions having the two-valued character of classical logic, i.e., which are either true or false for any values of the arguments. According to Reichenbach, the laws of quantum mechanics belong especially to this class of true-false propositions. The situation, as we see, is quite original: the laws of quantum mechanics are two-valued, but the reasonings in which they are included make use of a three-valued logic. On this ground Reichenbach thinks that logic is only a means of improving the language of science, but that it is not the reflection of general properties of some sphere of reality.

In Reichenbach's system there are the following tautologies (laws): R^1xx, $R^1xN^2N^2x$, $R^1xN^1N^1N^1x$, $R^1N^3xN^3N^3N^3x$, $R^1N^3xAN^1xN^1N^1x$, etc. The law of the excluded middle for diametrical negation is not preserved. For cyclical negation the law of the excluded fourth $AAxN^1xN^1N^1x$ and $R^1AxN^1N^1xN^3x$ is valid. The law of contradiction is preserved in such forms as: N^3KxN^3x, N^3KxN^1x, N^3KxN^2x. The law of De Morgan is preserved only in the forms $N^2Kxy = AN^2xN^2y$ and $N^2Axy = KN^2xN^2y$. The distributive laws are preserved as in two-valued logic. The law of contraposition is preserved in two forms: $R^1C^1N^2xyC^1N^2yx$ and $R^1C^2N^3xyC^2N^3yx$. The law of the definition of equivalence is preserved in the forms $R^2R^2xyKKC^2xyC^2yxKC^2N^2xN^2yC^2N^2yN^2x$ and $R^1R^1xyKC^1xyC^1yx$. The law of the definition of implication is preserved in the form $R^1C^2xyN^1N^2AN^3xy$. The *reductio ad absurdum* is preserved in the forms $C^1C^1xN^3xN^3x$ and $C^2C^2xN^3xN^3x$. For propositions taking only the values true *(1)* and false *(2)*, the law of the excluded middle is preserved for diametrical negation: if x is such a true-false proposition, then AxN^2x is a tautology.

Two propositions x and y are called complementary if they satisfy the relation (if with respect to them it is true that) $R^2AxN^1xN^1N^1y$. Because AxN^1x is true when x is true and when x is false, N^1N^1y must be true. This is the case when y is indeterminate. Thus x and y are complementary to one another if their relation satisfies the following condition: if x is true or x is false, then y is indeterminate. It can further be proved that, if x and y are complementary, then $R^2AyN^1yN^1N^1x$, i.e., the condition of complementarity is symmetrical for x and y. And this condition can also be formulated for three and more propositions. For example, for three it has the form: $R^2AxN^1xN^1N^1y$ and $R^2AyN^1yN^1N^1z$.

According to Reichenbach, the book of quantum phenomena is written in the language of three-valued logic. He considers it justifiable to speak about the truth and falsity of propositions only when it is possible to submit the propositions to a verification. If this is not possible, i.e., if the propositions can neither be verified nor falsified, then a third value – indeterminate – has to be given to them. Among such propositions are the propositions about unobservable objects. If the physicist adopts the ordinary methods of reasoning which are suitable for macrophenomena, then such an explanation of his experiments in the field of microphenomena leads to 'causal anomalies'. For example, a particle simultaneously exercises influences in places at which it does not occur itself; a field suddenly contracts into a point object; etc. The way out which Reichenbach sees lies in the admission of propositions about unobservable objects (in the given case: about microphenomena) into the class of indeterminate propositions.

We shall give an example of such an application of three-valued logic. Reichenbach uses the condition of complementarity for the description of facts concerning mutual limitation in the exactness of quantities measured (e.g. the relation of indeterminacy of Heisenberg): $C^2A[(e^1, t) = u]$ $N^1[(e^1, t) = u]$ $N^1N^1[(e^2, t) = v]$ which reads: if it is true or if it is false that the value e^1 at time t is u, then the proposition that the value e^2 at time t is v will be indeterminate.

As we see, Reichenbach bases his many-valued system on a very definite philosophical setting, with respect to which the following should be noted. Although Reichenbach's three-valued logic can be presented as a convenient means for reasoning about two-valued laws of quantum mechanics, this does not at all eliminate the question as to what causes the 'convenience' of this logic. It is evident that it must be connected with properties of the very reality under consideration and with the conditions of its knowledge. Otherwise, how would the 'inconvenience' of two-valued logic be explained in such cases?

Destouches-Février starts out from another philosophical setting and gives a different formal construction. For Destouches-Février, logical theory is a theory of being reflecting the general properties of the world. A given logical theory may be true for one part of the world and false for another part. Classical two-valued logic is true for the macrocosmos but not for the microcosmos. For the latter, instead of two-valued logic, the three-valued logic of complementarity is valid. For Reichenbach, on the other hand, logical theory does not pretend to be a true general theory of the world; it is only used for the removal of logical difficulties occurring without its application. (A comparison of the systems of Reichenbach and Destouches-Février from a formal point of view is made in TÖRNEBOHM.)

In the conception of Destouches-Février, as we see, the other aspect is stressed: namely, the fact that logic reflects some general properties of reality. But this conception is in a difficult position with respect to the following questions:

1) Many-valued logic is used even in the domain of macrophenomena; this means that it is not something specifically for reasoning only in the domain of microphysics.

2) In many cases a description by means of many-valued logic is possible, but not necessary. Even in quantum mechanics it does not seem to be an absolute constant necessity, considering that the further development of physics may well bring unexpected events. One should remember, especially in the present stage of the investigations, that one-sidedness in the solution of this kind of problem can give no positive results.

The system of Reichenbach has been introduced exclusively because it not only exemplifies the mutual relations between many-valued and two-valued logic as considered before and furthermore aims at a definite interpretation, but it also calls attention to the following important circumstance: a many-valued logic does not exclude two-valued propositions which only take the value truth or falsity. This is important from the point of view of the methodology of science, because many-valuedness is thus interpreted as a limitation which, under certain conditions of cognition, is put on reasoning with ordinary two-valued propositions.

The fact that many-valued logic includes two-valued propositions is easily explained: First of all, every n-valued construction $(n \geqslant 2)$ contains a class of functions which take only two values, though their arguments can take more than two values. Such functions are, for example, negation as defined by the Heyting matrix, alternative implication and alternative equivalence in Reichenbach's system, etc. Secondly, two-valued propositions can be constructed even with the help of many-valued functions. For instance, in the three-valued logic of Łukasiewicz Cxy, Kxy, Axy and Nx can take all three truth values 1, $\frac{1}{2}$ or 0, but by means of the symbols of the functions C, K, A and N, propositions can be constructed which take only the two truth values 1 or 0; so $CAxNxKxNx$, since $CA1N1K1N1 = CA10K10 = C10 = 0$, $CA0N0K0N0 = CA01K01 = C10 = 0$, $CA\frac{1}{2}N\frac{1}{2}K\frac{1}{2}N\frac{1}{2} = C\frac{1}{2}\,\frac{1}{2} = 1$.

There is even a third way: any given proposition constructed by means of symbols of many-valued functions can be subjected to limitation with respect of the truth values it can take. We simply determine for which values of the arguments this limitation holds; that is, under which conditions the proposition becomes two-valued. For example, if we let

$1, 2, \ldots, n$ be the truth values, $Cxy = max(1, 1-x + y)$, $Axy = max(x, y)$ and $Nx = x + 1$ (whereby $n + 1 = 1$); then $CAxNxNNy = 1$ if $x \geqslant y + 1$, and $CAxNxNNy = n$ if $y = n-3 + x$ or if $y = n-2 + x$. For $n = 3$ this proposition will have the value 2 only in the case where $x = 1$ and $y = 3$; the limitation in question can therefore be achieved by excluding this case from the number of possible ones.

§ 6. That many-valued logic can not only be used for the description of the language of quantum mechanics (or for the description of the phenomena of the microcosmos themselves) and that, therefore, it is not a 'privilege' of this sphere of reality and of the science describing it, is attested to by the work of Šestakov (cf. ŠESTAKOV 1946, 1953, 1960), in which he demonstrates the possibility of taking relay circuits as models of many-valued constructions. Thus in a model of a three-valued propositional calculus which consists of three-place switches controlled by three-place relays, the parameters of the circuit constructed from these switches and relays correspond (in a one-to-one correspondence) to three-valued propositions. As a matter of fact, the process here is more complicated: it is achieved by an arithmetical interpretation of the calculus and by the construction of the calculus itself in such a way that it can be easily modelled. And this is an achievement which is far from being trivial.

From the point of view of our subject, the works of Šestakov are also interesting with respect to the solution of the question of the mutual relations between two-valued and n-valued logic and of the question of the mutual relations between different many-valued constructions.

In all of the cases of the construction of many-valued logical systems considered above, the construction itself was accomplished by means of two-valued logic (i.e. the language by means of which the construction of n-valued systems was accomplished, was itself subject to ordinary two-valued logic). As a matter of fact, Rosser and Turquette are of the opinion that two-valued logic is sufficient but not necessary for the construction of many-valued systems – a question which needs special consideration. However, the fact remains: up to now in all cases known the matter stands thus.

Let $1, 2, \ldots, n$ be truth values. The proposition 'proposition x has the truth value i' (abbreviated: xi) is two-valued, i.e., either proposition x has in fact the truth value *i* or it does not have it (it has another of the possible values *1*, *2*,..., *n*). Every proposition of *n*-valued logic takes only one of the values *1*, *2*,..., *n*. This conception either explicitly or implicitly lies at the basis of every construction of *n*-valued logic, and is presupposed in the construction of truth tables, of the equations equivalent to them (the formula xi means the same as the formula $x = i$), etc. Considering both what has been said and also the circumstance that propositions like xi are at the same time propositions of *n*-valued logic, Šestakov has shown that any function of *n*-valued logic can be represented by functions of two-valued logic.

Šestakov constructs a three-valued calculus taking as a basis Webb's function (cf. WEBB 1935, 1936) in its three-valued variant. If *1* is considered as falsity, *2* as meaninglessness or indeterminacy, *3* as truth, then Webb's function is defined by the table:

x \ y	1	2	3
1	2	3	1
2	3	3	1
3	1	1	1

Webb's function proves to be a generalization of the two-valued Sheffer function, since it has the characteristic that it alone is sufficient for the construction (the definition) of all possible functions of *n*-valued logic (in the three-valued case; of all functions of three-valued logic). Defining several other functions by means of this function, Šestakov has shown that the calculus of Bočvar is entirely contained in the three-valued construction of Kleene (cf. ŠESTAKOV 1960). Thereby 'meaningless' in the sense of Bočvar corresponds to 'undefined' in the sense of Kleene. (But in order to show that the calculus of Kleene is stronger than that of Bočvar, it is not necessary to take into consideration the interpretation of the third value). Because Webb's function (we symbolize it Wxy) is taken as basic, the functions F^1x, F^2x, Nx, Axy and Vx (for the meaning of the symbols N, A and V cf. § 4) can be defined respectively as $F^1x = Wxx$, $F^2x = WF^1xF^1x$, $Nx = WWF^1xF^2xF^2WxF^2x$, $Axy = F^2Wxy$, $Vx = WF^1xF^2x$, and the calculus constructed by Šestakov proves to be functionally complete (it can comprehend all possible functions of the three-valued propositional calculus). It is obvious that by means of the definitions $Kxy = NANxNy$, $Cxy = ANxy$, etc., the other functions of the system of Bočvar are definable on the basis of Webb's function.

§ 7. There is one further source of many-valued logic: the systems of strict implication (Lewis, Ackermann ('*strenge Implikation*'), cf. ACKERMANN, VON WRIGHT, ROSSER and TURQUETTE). These systems resulted from recognizing that under a strict formalization of logical consequences classical propositional logic (constructed by means of two-valued matrices or also equivalently as complete axiom systems or generally as any kind of equivalent formal system) is not absolute. For example, the tautologies (i.e., the provable formulas) of two-valued logic *CxCNxy*, *CyCxy*, *CKxNxy*, *CxAyNy*, etc. are considered as 'paradoxical' in the sense that they cannot be accepted as rules of deduction if the symbol *C* is interpreted as the symbol for deduction (reading, e.g., the first tautology as 'from *x* it can be deduced that from not-*x* any proposition can be deduced', or in a more customary way as 'from a false proposition anything follows').

The systems of strict implication (both Lewis's and Ackermann's) set themselves the task of excluding such formulas from the number of valid ones. Naturally, when considering many of the problems concerning the properties of these systems, one cannot use the construction by two-valued matrices as an interpretation. A many-valued construction seems appropriate here, because it offers considerable room for definitions of functions and of tautologies where the exclusion of undesirable formulas from the number of tautologies (i.e. of provable formulas) is relatively easy.

In the logic of strict implication several attempts to use many-valued interpretations have been made. For example, ACKERMANN has recourse to the following six-valued interpretation. The truth values are symbolized by *0*, *1*, *2*, *3*, *4*, *5*. Value *2* signifies absurdity. As proved formulas (tautologies) those formulas are taken which take any one of the values *3*, *4*, or *5* for any values of the arguments whatsoever. As we see, the tautologies can take here several values instead of just a single one as in two-valued logic. Negation is defined in the same way as the second negation in Post, i.e. $Nx = 5-x$ (in this case one does not have to add *1*, because the values start here with *0*). The definition of implication runs as follows (using the notation of Łukasiewicz):

1) $Cxy = 3$, if $x = y$, if $x = 0$, if $y = 5$, if $x = 1$ and $y = 2$, if $x=1$ and $y = 4$, if $x = 3$ and $y = 4$;

2) $Cxy = 0$ in all other cases.

It can be shown that several tautologies of two-valued logic are not tautologies in Ackermann's system. For example, $CxCyx = 0$, if $x = y = 2$; and $C2C22 = C23 = 0$. Also $CxCNxy$, $CKxNxy$, etc., are not tautologies here. But among the tautologies belong, e.g. the axioms of the system of strict implication of Ackermann. Take, say, $CCx2Nx$. By substitution we get $CC02N0 = C3N0 = C35 = 3$, $CC12N1 = C3N1 = C34 = 3$, and so on for the remaining cases. There are also other interpretations of this system of strict implication (cf., e.g., ROSSER and TURQUETTE).

In ZINOV'EV 1961a questions concerning the implications of the Lewis-Ackermann type are considered in more detail. In particular, it is shown there that a many-valued or generally a functional interpretation of this type of implication is only one of the heuristic means possible in this domain of logic.

CHAPTER THREE

GENERAL QUESTIONS CONCERNING MANY-VALUED LOGIC

§ 1. Above we have seen that widely differing variants of many-valued logic are possible. On the basis of the experience accumulated in their constructions, some general principles of construction can be formulated, as it has been done, e.g. in ROSSER and TURQUETTE and SUSZKO. It should be noted that these general principles can, without affecting their content, be stated in very different terminologies; and, of course, this can be done in a way which is much more strict and detailed than the one we will choose below.

Let us turn, first of all, to the constructions by means of truth tables or of functions of truth values where the many-valued character is plain from the very beginning. Furthermore, this approach is natural not only from a logical but also from an historical point of view: the many-valued calculi originated in functional forms.

Actually, in the history of contemporary logic the axiomatic constructions are older than the functional ones: Peirce introduced the method of matrices only in 1885, whereas already in 1879 Frege had published the first axiomatization of the propositional calculus. But at that time all logical constructions were based on the conviction that the propositions were essentially two-valued, and the mutual relations between different types of construction did not give rise to philosophical questions of principle.

The functional constructions are based on the following principles (at least on these). Any finite sequence of symbols, where the latter are taken from a given finite or denumerably infinite set of symbols, is by definition a formula, and from the set of formulas, the formulas of some non-empty subset are chosen as propositions (or formulas) of the given logic under construction. For example, the following formulas can be chosen as propositions of some given logic:

1) $x, y, z\ldots$ (small Latin letters) with or without indices are propositions;

2) if x is a proposition, then Nx and Tx are propositions;

3) if x and y are propositions, then Cxy is a proposition;

4) further propositions can be defined by means of the ones already mentioned.

The positive integers $1,\ldots, n$, where n is an integer and $n \geqslant 2$, are called truth values. If the numbers from 0 to $n-1$, from 0 to 1, or some other numbered symbols, say $t^0, t^1,\ldots,$ are taken as truth values, then it is not difficult to reduce them to $1,\ldots, n$ by counting them over by means of $1,\ldots, n$ (by correlating them to $1,\ldots, n$).

It is assumed that the propositions take one of the values $1, \ldots, n$. Further, some infinite number (normally a small one) of propositional functions $F^1(x^1,\ldots, x^k)$, where $k \geqslant 1$, are considered (assumed, given). The distinction between propositional functions and truth functions made in ROSSER and TURQUETTE is in our case not essential. These propositional functions are complex propositions, the truth value of which is defined as depending on the truth values of the propositions composing them (of the arguments). The definitions are given by means of matrices or equations of truth values. Thus, $F^1(x, y)$, $F^2(x)$ and $F^3(x)$ (respectively Cxy, Nx and Tx) are defined by the equations $F^1(x, y) = max(1, y-x+1)$, $F^2(x) = n-x + 1$, $F^3(x) = 2$. In ROSSER and TURQUETTE and also in SŁUPECKI 1930 these functions are taken as basic for three-valued logic. In SŁUPECKI Cxy and Nx are defined as in Łukasiewicz and Tx is called the Słupecki function.

The constructed functions are the basic (given) functions F^i and the functions obtained from them by way of combination and substitution. Examples of constructed functions are F^1, F^2 and F^3, but also functions of the type $NCxTy$ $(F^2F^1xF^3y)$, $NCCxyNz$, etc.

In the case of functional constructions, the problem of consistency is not of primary importance, because the consistency is assured by the very methods of construction: if some truth value is assigned to a proposition, then it is not at the same time denied to it. This means that the conditions of the construction exclude the cases where $x = i = k$ (i and k being different numbers) or cases where one and the same compartment of a matrix has two different values i and k (which can be read 'x has the value i and at the same time the value k', i.e. not i). This circumstance is also important in another respect: the consistency of the functional constructions is assured by the fact that the propositions of the type $x = i$ are two-valued and that propositions of the type 'x has and at the same time does not have the truth value i' (this is a more general formulation of the proposition given above) are excluded. Also the problem of the independence of the basic functions is not of primary importance: if it can in some way be shown that one of the basic functions can be defined in terms of the others, then this can only lead to a reduction in the number of basic functions. Thus Axy and Kxy can be defined by means of Cxy and Nx. But the reduction of some functions to others is important for the choice of the simplest axiom system corresponding to the given functional constructions. Thus, in the axiom systems of Wajsberg-Słupecki and Rosser and Turquette, the symbols A and K are not used, because they are definable in terms of C and N.

For n values the number of possible functions of m arguments is equal to n^{n^m}. Therefore the problem of the functional completeness of the calculus (i.e. the question whether the totality of all functions which can be constructed in the calculus does or does not contain all possible truth functions) is of special interest. Above we have already met some examples of functionally complete many-valued logics of proposition: the systems of Post, Webb and the one of Kleene-Bočvar in the construction of Šestakov. And we will give some more examples which show that the vast number of possible functions is no hindrance for the construction of complete systems. One must remember, of course, that the question of completeness and the question of the convenience of a construction for some purposes are questions of an essentially different order. Thus the system of Webb has only one basic function, but this calculus proves to be somewhat cumbersome.

Słupecki has shown that the system of Łukasiewicz, which takes the functions Nx and Cxy as basic, is not functionally complete. In particular, it is impossible to define the function T, for which $Tx = \frac{1}{2}$ for any value of x, in terms of the operators C and N alone. Also, in n-valued logic (with the truth values $1, 2, \ldots, n$), where $Tx = 2$, T is not reducible to C and N. It might be noted in passing that this function has no analog in two-valued logic. Since it is not the only one of its kind (e.g. $T^1x = 2, T^2x = 3, \ldots, T^{n-2}x = n-1$ are all possible), we can speak of a whole class of functions which have no analog in two-valued logic. In this sense many-valued systems cannot be reduced to two-valued ones or simply to a generalization thereof, but appear also as somehow existing alongside to them. And on the other hand, interpreting the operators of two-valued logic as operators of n-valued logic does not always give a complete n-valued calculus.

Słupecki has further shown that a logical system based on the functions

Nx, *Cxy* and *Tx* is functionally complete. That is to say, Słupecki proved this for three-valued logic. In ROSSER and TURQUETTE it is shown that a logical system based on the functions F^1x, F^2xy and F^3x (*Nx*, *Cxy* and *Tx* respectively) is functionally complete where these functions are characterized as follows: if $1, 2, \ldots, n$ are the truth values, then $F^1x = n-x+1$, $F^2xy = max(1, y-x+1)$, $F^3x = 2$. Axiom systems corresponding to these functional constructions will be given below.

Further, from the numbers $1, 2, \ldots, n$, an s such that $1 \leqslant s < n$, is chosen. The truth values $1, \ldots, s$ are called 'designated' and the values $s+1, \ldots, n$ are called 'undesignated'. Propositions which have a designated truth value are called 'asserted', and those having an undesignated one are called 'denied'. Moreover, the choice of s is relative (the 'logical coordinates' are relative). As we shall see, there is an analogy here with two-valued logic, namely with truth and falsity, and also with assertion and negation. Only it is possible here for a proposition to be asserted (or denied) for two or more values (cf., e.g., Ackermann). But there is the particular case where $n = 2$ and $s = 1$. The choice of s has no influence on the internal structure of a functional calculus. It only determines the classes of asserted and denied propositions, something which is particularly important for axiomatic constructions. Although the number of designated (or undesignated) truth values can be greater than one, i.e., although within the class of asserted (or denied) propositions a further division can be made, the principles of two-valued logic remain fully valid in the sense that a proposition is either asserted or denied (no proposition can at the same time be both asserted and denied). The choice of n, F^i and s determines the character of the calculus.

§ 2. The fundamental task of functional constructions consists in reducing the investigation of complex functions to elementary ones and in finding the criteria of completeness for the calculi based on these elementary functions. From this point of view, the works of Słupecki, Rosser and Turquette, Jablonskij and A. V. Kuznecov are of interest (cf. JABLONSKIJ, where references to the work of Kuznecov can be found). A full characterization of these works is impossible here because of their highly technical character. We confine ourselves to some brief remarks.

In JABLONSKIJ the set (say V) of all functions $F(x^1, \ldots, x^n)$ of n arguments is considered, where arguments and functions take one of the values $0, 1, \ldots, k-1$. Further, the functions constructed on the basis of the system (say V^*) of some functions $F^1, \ldots, F^s$ are considered, where $F^1, \ldots, F^s$ are selected from the set V. A function obtained by replacing variables (whereby functions of the system V^* can also be substituted for the variables $x^1, x^2, \ldots, x^n$) is called a 'superposition' of the functions of system V^*. A set (say W) of functions selected from the set V is called 'functionally closed', if not only the functions $F^1, \ldots, F^s$ but also any of their superpositions belong to it. A system of functions from the set W is called 'complete in W', if every function from W is a superposition of functions of this system.

These definitions express in another form and in another terminology what we have been discussing in the previous section: some functions from the set of all possible functions are selected as basic, a way of forming new functions (by replacing variables) is given, and the system is called complete if the set of functions which can be constructed in this system coincides with the set of all possible functions (for n arguments and k values).

In JABLONSKIJ a system of theorems concerning functional completeness is developed and an exhaustive treatment of these problems is given in the case of three-valued logic. In particular, it is shown that the following system of functions is functionally complete:

1) $F^1x = 0$, $F^2x = 1$; …; $F^kx = k-1$; $Q^1(x, y) = max(x, y)$; $Q^2(x, y) = min(x, y)$; $J^ix = k-1$ for $x = i$, and $J^ix = 0$ for $x \neq i$.

Also the following systems are complete:

2) $F^1(x, y) = max(x, y)$; $F^2(x) = x + 1 (mod\ k)$.

3) $F(x, y) = [max(x, y) + 1]\ (mod\ k)$, where F is Webb's function, the analog to Sheffer's function.

§ 3. The following is a further example of a functionally complete three-valued construction. Let 1, 2 and 3 be the truth values. We take the following functions as basic:

1) $Axy = min(x, y)$;

2) $Bxy = 2$, if $x = y = 2$; $Bxy = 3$, if $x = y = 3$; $Bxy = 1$ in all the other cases; Bxy is a special generalization of two-valued disjunction, different from Axy;

3) $Nx = 4-x$;

4) $Mx = x + 1$, if $x = 1$ or $x = 2$; $Mx = 1$, if $x = 3$. By means of

these functions Cxy with the meaning $Cxy = max(1, 1-x+y)$, and Tx with the meaning $Tx = 2$ can be defined as $Cxy = AANxyMMBxy$, $Tx = MBxMx$. As the construction C–N–T is functionally complete, the same holds also for the construction A–B–M–N. The definitions mentioned can easily be verified: $T1 = MB1M1 = MB12 = M1 = 2$, $T2 = MB2M2 = MB23 = M1 = 2$, $T3 = MB3M3 = MB31 = M1 = 2$. Similarly a verification for C can be obtained. In fact, the matrix for $ANxy$ differs from the matrix of Cxy only in the case $x = y = 2$ where $Cxy = 1$ and $ANxy = 2$; $MMB22 = MM2 = 1$, $MMB33 = MM3 = 2$ and in all other cases $MMBxy$ has the value 3; therefore if $x = y = 2$, then $AANxyMMBxy = 1$, and in all other cases the value $ANxy$ is retained.

The example with the system A–B–M–N has been introduced not only in order to show that different selections of basic functions are possible but also in order to call attention to the following important circumstance. The functions B and M can be considered as special generalizations of two-valued disjunction and negation differing from the corresponding generalizations A and N. This can easily be seen: for $n = 2$, $M1 = 2$, $M2 = 1$, $B11 = 1$, $B12 = 1$, $B21 = 1$, $B22 = 2$.

Thus different generalizations of two-valued logic can occur together in one and the same many-valued logical construction, so that analogs of the two-valued laws can appear in different planes. In the given example, $BxMx$ is a generalization of the law of the excluded middle. Thus $BxMx = 1$, if $x = 1$, $x = 2$ or $x = 3$; i.e., if those propositions which always have the truth value 1 are accepted as laws (tautologies), then $BxMx$ is a law in three-valued logic. The same is not true for $AxNx$: $A2N2 = A22 = 2$. If, therefore, one speaks of the preservation or non-preservation of some law of two-valued logic in a given many-valued construction, then one must indicate exactly which generalization of two-valued functions one has in mind.

This remark is very important, because among the possible functions of any n-valued propositional calculus, at least one pair of functions A^* and N^* can be found such that they are generalizations of the two-valued functions A and N respectively, and A^*xN^*x is a law for any definition of 'law' (i.e., for any choice of truth values under which propositions are considered to be assertions). Such a generalization of A and N must fulfil the following necessary condition: if the value a $(1 \leqslant a \leqslant n)$ corresponds to truth and the value b to falsity, then $N^*a = b$, $N^*b = a$, $A^*aa = A^*ab = A^*ba = a$ and $A^*bb = b$. Thus, in the sense now being considered, the law of the excluded middle remains valid in any n-valued propositional calculus (namely in the form A^*xN^*x). With respect to the other basic laws of two-valued logic the same holds true (but with corresponding modifications in the definitions of the functions). What has been said above is sufficient to raise doubts concerning the expediency of using the expression 'law of the excluded middle' in application to laws of logical calculi (we will return to this below).

§ 4. A survey of all possible functions of the n-valued logic of propositions by means of a very simple algorithm has been considered in ZINOV'EV 1961b. This paper is reproduced here because it contains a proof of the completeness of a system with certain basic functions which, because of its simplicity, is quite suitable for a work of philosophical scope.

Atomic propositions will be designated by the symbols $x^1, x^2, \ldots, x^m$, and truth values (values of arguments and functions) by the symbols $1, 2, \ldots, n$. The set of values $1, \ldots, n$ we assume as given. We take as basic the functions K, B and M which we define in the following way:

1) $K(x^1, \ldots, x^m) = max(x^1, \ldots, x^m)$ where $m \geqslant 2$;

2) $M^i x = x + i$ where $0 \leqslant i \leqslant n$, and if $x + i > n$ then $M^i x = x + i - n$; (in general, whenever an addition of truth values will give number α and $\alpha > n$, then we will replace α by $\alpha - n$; and whenever a subtraction will give α and $\alpha < 1$, then we will replace α by $\alpha + n$);

3) $B(x^1, \ldots, x^m) = i$, if all arguments $x^1, \ldots, x^m$ have the value i $(x^1 = \ldots = x^m = i)$ where $1 \leqslant i \leqslant n$; $B(x^1, \ldots, x^m) = 1$ in all the other cases (if the values of at least two of the arguments $x^1, \ldots, x^m$ are different). Obviously $B(x, M^1x) = 1$, because $x \neq x + 1$, $M^iB(x, M^1x) = i + 1$, $M^iM^kx = M^{i+k}x$. Below we shall show that on the basis of the functions K, B and M a standardized method can be described which gives a survey (the definition, construction) of all possible functions of the n-valued calculus of propositions. Instead of M^ix we can take $Mx = x + 1$. In this case M^i means M taken i times (if $i = 0$, then there is no M in front of x) and is simply an abbreviation.

We introduce the function W, indicating how to define it by means of B and M. This function will have the following properties: for any combination of values of the arguments, it will have the value 1, except for one single combination for which it will have the value $1\ (i \leqslant i \leqslant n)$, where i depends on the choice of certain constants; thus the exceptional combination of values of the arguments for which the function will have the value i can also be any of the possible ones, depending on the choice of these constants.

For one argument, Wx can be defined:
$Wx = B(M^k x, M^{i-1+k} B(x, M^1 x))$ where $0 \leqslant k \leqslant n-1, 1 \leqslant i \leqslant n, 1 \leqslant i+k \leqslant n$. Thus i and k are in any case constants chosen by us between the limits mentioned. Therefore, this definition of Wx, and likewise the later definition of $W(x^1,\ldots, x^m)$, is the definition of a whole class of functions. Below we will indicate how to distinguish the elements of this class.

It can easily be verified that the definition of Wx satisfies the demands formulated above: according to the determination of B and M and the assertion $B(x, M^1 x) = 1$ we get $Wx = B((x + k), (i + k))$; if $x \neq i$, then $Wx = 1$; if $x = i$, then $Wx = x + k = i + k$. Thus, only in a single case can Wx have a value different from 1 (however as a particular case it may be that $x + k = i + k = 1$), namely if $x = i$. Thus, depending on the choice of i and k, Wx can vary.

For two and more arguments $x^1,\ldots, x^m$, the function $W(x^1,\ldots, x^m)$ can be constructed in the following way.

First step: we choose some combination of values of the arguments (the conditions limiting this choice will become clear in the following; but as far as successively all possible combinations will be open to choice, it is sufficient to speak here simply about a few of them).

Second step: we compare the values of the arguments; if they are not equal, then in front of every argument we put M with corresponding indices (the index can be 0) so that all values of the arguments of the given combination become equal to $j\ (1 \leqslant j \leqslant n)$. Depending on the character of the function to be defined, there are various possible ways of making the values equal.

Third step: we put $M^k\ (0 \leqslant k \leqslant n-1)$ in front of every argument; thus we get for W and the given combination of values $x^1,\ldots, x^m$ any value $k + j$ that we want from the numbers $1,\ldots, n$.

Fourth step: we form the function B, putting as arguments $x^1,\ldots, x^m$, together with the M symbols we have written in front of them according to the second and third step. We designate this function by the symbol I.

It will have the general form: $B(M^{r^1+k}x^1, M^{r^2+k}x^2, \ldots, M^{r^m+k}x^m)$, where $r^1 + x^1 = r^2 + x^2 = \ldots = r^m + x^m = j$.

Fifth step: we form the function $M^{j+k-1}B(x, M^1x)$, where x is any one of the arguments $x^1, \ldots, x^m$. We designate it by the symbol *II*.

Sixth step: we form the function $B(I, II)$.

The function $B(I, II)$ obtained is $W(x^1, \ldots, x^m)$, since *I* and *II* have the same value $k + j$ only in one case, and in all other cases their values are different. This means that $B(I, II)$ has for only one single combination of values of the arguments the value $k + j$ (particular case: $k + j = 1$) and for all others the value *1*. The selection of k and j can be such that $k + j$ will be equal to any one of the values $1, \ldots, n$, and the place where $B(I, II)$ will have this value depends on our choice (i.e. on the position of M in front of the arguments $x^1, \ldots, x^m$).

Let us illustrate this procedure for the example $W(x^1, x^2)$. Let $x^1 = a$ and $x^2 = b$, where $1 \leqslant a \leqslant n$ and $1 \leqslant b \leqslant n$, be the combination of values chosen. $W(x^1, x^2)$ has the form:
$B(B(M^{r+k}x^1, M^{s+k}x^2), M^{j-1+k} B(x, M^1x))$ where $1 \leqslant r + k \leqslant n$, $1 \leqslant s + k \leqslant n$, $j = r + a = s + b$. It can easily be verified that this is $W(x^1, x^2)$: this expression has the same value as $B(B((r + k + x^1), (s + k + x^2)), (j + k))$; if $x^1 = a$ and $x^2 = b$ we get $B(B((j + k), (j + k)) = j + k$; in all other cases it is equal to *1*. Thus $1 \leqslant j + k \leqslant n$, and the choice of the combination a and b is arbitrary.

We come to the next stage: we see how it is possible, by means of K, B and M and also of the function W defined on their basis to define all possible n-valued functions. We consider first the functions of one argument x. We take Fx, characterized by $Fx = a^i$ (where $1 \leqslant a^i \leqslant n$) if $x = i$ (where $1 \leqslant i \leqslant n$). The a^i $(a^1, \ldots, a^n)$ are subject to only one condition: they must be values from the set $1, \ldots, n$. Therefore Fx is an arbitrary function of x. Let $W_1x = a^1$, if $x = 1$; $W_2x = a^2$, if $x = 2; \ldots$; $W_nx = a^n$, if $x = n$. Obviously, $Fx = K(W_1x, \ldots, W_nx)$.

The situation is similar for two or more arguments. Let all the possible combinations of the values $x^1, \ldots, x^m$ be numbered by $1^*, 2^*, \ldots, n^*, \ldots, n^{m*}$. We take an arbitrary function $F(x^1, \ldots, x^m)$ and set up the correspondence: to combination 1^* corresponds the value a^1 of this function, to combination 2^* the value $a^2, \ldots$, to combination n^{m*} the value a^{n^m}. We also number W in the following way: $W_1(x^1, \ldots, x^m) = a^1$, if 1^*; $W_2(x^1, \ldots, x^m) = a^2$, if $2^*; \ldots$; $W_{n^m}(x^1, \ldots, x^m) = a^{n^m}$, if n^{m*}.

Obviously, $F(x^1,\ldots, x^m) = K(W_1(x^1,\ldots, x^m),\ldots, W_{n^m}(x^1,\ldots, x^m))$.

The method set forth here can be modified somewhat. Let Pa be an arbitrary function which has in place of a any finite number of arguments. P can either be added to the number of basic functions, or it may already be one of them, or it can be constructed by means of them. Let Qa be any other function which should be constructed (defined, described) by means of K, B, M and P. Let the combination of values of the arguments 1^* correspond to the value u^1 of the function Pa and to the value v^1 of the function Qa, the combination 2^* to u^2 of Pa and to v^2 of $Qa,\ldots,$ the combination n^{m*} to u^{n^m} of Pa and to v^{n^m} of Qa. Let u^i be any value from $u^1, u^2,\ldots, u^{n^m}$, and v^i correspondingly any value from $v^1, v^2,\ldots, v^{n^m}$. We will show that an F^ia can be constructed, which will differ from Pa only by the fact that for the combination of values i^* the function F^ia will have the value v^i, and not u^i as Pa.

This is accomplished in the following way:

First step: we construct M^kPa where $k = n-u^i + 1$. This means that for the combination of values i^*, $M^kPa = 1$.

Second step: we construct W_ia such that $W_ia = t$, if $Pa = u^i$.

Third step: we construct $K(M^kPa, W_ia)$. This function will differ from M^kPa only by the fact that for $M^kPa = 1$ it will have the truth value t (particular case: $t = 1$).

Fourth step: we form the function $MjK(M^kPa, W_ia)$ where $j + t = v^i$. As we want to obtain Qa, we put $j = n-k$. Obviously, then $n-k + t = v^i$. This is the formula for the choice of t: $t = v^i + k-n = v^i + n-u^i + 1-n =$ $= v^i + 1-u^i$.

The function $MjK(M^kPa, W_ia)$ constructed is F^ia. Since F^ia is a function constructed for an arbitrary combination of values of the arguments i^*, it can obviously be constructed for the combination 1^*. We symbolize it by F^1a. Now we take F^1a as given and by the same method we construct the function $F^{12}a$, which will differ from F^1a only by the fact that for the combination 2^* it will have the value v^2 and not u^2. By passing through all combinations in this way, we get $F^{12\ldots n^m}a$, which is Qa.

§ 5. The definitions of functions of the logic of propositions have the following structure. For functions of one argument the definition can be represented as a set of n (where n is the number of truth values) conditional propositions of the type: if x has the truth value i, then $F(x)$ has the truth value k (where i and k are values from the set of given values; i and k may be identical). For functions of two or more arguments the definition can be represented as a set of n^m (where m is the number of arguments) conditional propositions in whose antecedents occur all possible combinations of truth values of the arguments. For example, the definition of $F(x, y)$ consists of n^2 conditional propositions of the type: if x has the truth value i and y has the truth value k, then $F(x, y)$ has the truth value j.

The symbols $F(x)$, $F(x, y)$, Axy, $K(x, y, z)$, $F(x^1, \ldots, x^m)$, etc., indicate on what arguments the truth values of some proposition depend and what kind of dependence it is. Moreover, these symbols also designate the propositions themselves, whose truth values are functions of the arguments occurring in the symbols. This way of designation suggests that $F(x)$, $F(x, y)$, Axy, $K(x^1, \ldots, x^m)$, etc., are propositions constructed respectively from x, y, $x^1, \ldots, x^m$. Thus Axy represents a proposition consisting of x and y. This, however, is only one of the possible interpretations of the functions of propositional logic, whereby the latter are considered as complex propositions. This interpretation is so widespread and so common that it seems connected with the definitions in a necessary way. But in fact this is not the case. Below we will show another interpretation. Here we simply want to say that the definitions indicate only the dependence of the truth values of some propositions on the truth values of others. And the symbols $F(x)$, $F(x, y)$, Axy, $F(x^1, \ldots, x^m)$, etc., signify according to the definitions only the following: $F(x)$ is a proposition, the truth value of which depends in the way F on the truth value of the proposition x; similarly for the other functions. The symbolism under consideration is, of course, convenient. But if all elements of the definitions are to be made explicit, then a symbolism, say of the following type, would be more accurate: $(y)F(x)$, $(y)F(x^1, x^2)$, $(y)Ax^1x^2$, $(y)K(x^1, \ldots, x^m)$ etc. (proposition y is a truth function of the type F of proposition x; similarly for the other symbols).

Some philosophers call attention to the possibility of reading the definitions in the opposite direction, i.e. from the functions to the arguments. Concerning this, it is sufficient to remark the following: if the definition of some function is given where in the antecedents of the conditional propositions all possible combinations of truth values of arguments are enumerated, then it is not difficult on the basis of the truth value of the function to estimate the truth values of the arguments. For example, if $Kxy = 3$ in three-valued logic, then either $x = 3$ and $y = 1$, or $x = 3$ and $y = 2$, or $x = 3$ and $y = 3$, or $x = 2$ and $y = 3$, or $x = 1$ *and* $y = 3$. It is true that often different possibilities excluding one another have to be enumerated in the consequent. But this may also hold for the transition from the arguments to the functions. Special attention should be paid to this latter case.

The functions of propositional logic can be divided into two groups. To the first group belong the functions characterized by the following property: to every combination of truth values of the arguments corresponds one and only one definite truth value of the function. To the second group belong the functions characterized by the property: to at least one combination of truth values of the arguments corresponds one of two or more truth values of the function (the function can take any of two or more truth values for one and the same combination of truth values of the arguments). The functions of the first group we call 'singular', those of the second group 'plural'.

We shall give some examples of plural functions. In five-valued logic (truth values: $1, 2, 3, 4, 5$) the following functions are possible:

1) if $x = 1$, then $F^1(x) = 1$; if $x = 2$, then $2 \leqslant F^1(x) \leqslant 4$; in all the other cases $F^1(x) = 5$.

2) if $x = 1$ and $y = 1$, then $1 \leqslant F^2(x, y) \leqslant 3$; if $x = 1$ and $y = 2$, then $3 \leqslant F^2(x, y) \leqslant 5$; in all the other cases $F^2(x, y) = 4$.

In n-valued logic (truth values: $1, \ldots, n$; $n \geqslant 2$) the following functions are possible:

1) if $1 \leqslant x \leqslant s$ $(1 \leqslant s \leqslant n-1)$, then $s + 1 \leqslant Q^1(x) \leqslant n$; if $s + 1 \leqslant x \leqslant n$, then $1 \leqslant Q^1(x) \leqslant s$.

2) if $x \leqslant y$, then $1 \leqslant Q^2(x, y) \leqslant s$; if $x > y$, then $s + 1 \leqslant Q^2(x, y) \leqslant n$.

The definitions of plural functions can be considered as definitions of sets of two or more singular functions. Thus $F^1(x)$ covers three singular functions, all similar except for the truth values for $x = 2$: $F^{11}(2) = 2$, $F^{12}(2) = 2$ and $F^{13}(2) = 4$. The set $F^2(x, y)$ contains nine singular functions. The function $Q^1(x)$ is a set of $(n-s)^s s^{(n-s)}$ singular functions.

Let $F(\alpha)$, where α is a non-empty set of arguments, be a plural function covering exactly m singular functions $F^1(\alpha), \ldots, F^m(\alpha)$ (for a given combination of truth values of the arguments the truth value $F^1(\alpha)$ is one of the truth values admissible according to the definition for $F(\alpha)$; similarly for every combination of truth values of the arguments; and in the same way for each of the functions $F^1(\alpha), \ldots, F^m(\alpha)$), such that each of the functions $F^1(\alpha), \ldots, F^m(\alpha)$ satisfies the definition of $F(\alpha)$ and no other singular function satisfies it. Then the definition of $F(\alpha)$ will be equivalent with the following definition:

1) each of the functions $F^1(\alpha), \ldots, F^m(\alpha)$ is defined;

2) $F(\alpha)$ is defined in the form '$F^1(\alpha)$ is $F(\alpha)$;...; $F^m(\alpha)$ is $F(\alpha)$; no other function is $F(\alpha)$'.

Normally the definitions are given in an abbreviated form, because such a full definition would be somewhat cumbersome and at times even impossible (in the case of very great or even infinite sets of truth values of the functions for given combinations of truth values of the arguments). But in a theoretical investigation of the properties of $F(\alpha)$ it is useful to present them in such a full form, because then their nature comes out clearly and some of their properties can be distinguished exactly.

Any two different singular functions which take the same arguments can be generalized to a plural function. Let, for some combination of truth values of the arguments α, the function $R^1(\alpha)$ have the truth value k, and the function $R^2(\alpha)$ the truth value j; we introduce $R(\alpha)$ by the principle 'for the given combination of truth values of the arguments, $R(\alpha)$ has one of the truth values k and j; doing the same for every combination of truth values of the arguments, we get the full definition of $R(\alpha)$'. Similarly any three, four, etc., singular functions which take the same arguments can be generalized to a plural function.

The definition of a plural function consists, first of all, in a generalization of several singular functions (in the introduction of a general symbol for two or more singular functions) and, secondly, in the definition of every one of these singular functions (or in giving a rule according to which this can be done in principle). Because a plural function is a generalization of two or more singular functions, it is impossible to define it by a combination of singular functions (by a formula of the Hilbert type): every formula constructed with propositional variables and symbols of singular functions is a singular function. Therefore the problem of completeness has a positive solution only for singular functions, but not for plural ones.

The converse, i.e., the definition in the form of a formula of the Hilbert type of a singular function in terms of plural functions, presents no problem. Let us give an example. Let $1 \leqslant F(x, y) \leqslant 2$, if $x = y = 1$, $F(x, y) = 1$ if $x = 3$ and $y = 6$, and $F(x, y) = 4$ in all the other cases; $3 \leqslant Q(x, y) \leqslant 5$, if $x = y = 3$, and $Q(x, y) = 6$ in all the other cases. We take $F(F(x, y), Q(x, y))$. For this formula we have the following possibilities: $F(1, 6) = 4$, $F(2, 6) = 4$, $F(3, 6) = 4$, $F(4, 3) = 4$, $F(4, 4) = 4$, $F(4, 5) = 4$ and $F(4, 6) = 1$. The case $F(1, 3)$ is excluded since it is impossible that $Q(x, y) = 3$ for $F(x, y) = 1$. Analogously, the cases $F(1, 4)$, $F(1, 5)$, $F(2, 3)$, $F(2, 4)$, $F(2, 5)$, $F(3, 3)$, $F(3, 4)$ and $F(3, 5)$ are all excluded. Consideration of the cases where $F(F(x, y), Q(x, y)) = 4$ shows that this holds in the following cases: (1) $x = y = 1$; (2) $x = y = 3$; (3) $x = 4$ and $y = 6$. In the remaining cases $F(F(x, y), Q(x, y)) = 1$.

If for some combinations of truth values of the arguments the truth values of the functions are not indicated, then we get incomplete definitions. In using such functions it is possible to consider them as plural functions with only this difference: that informally we have in mind our lack of knowledge of the value of the functions for these combinations truth values of the arguments (that we have in mind some of the n^m functions, where m is the number of combinations of truth values of the arguments for which the value of the function is not indicated).

§ 6. Let us make some general considerations concerning negation in many-valued constructions. If any function Fx of one argument, such that for at least one truth value $Fx \neq x$, is taken as a negation of x, then for n values n^n-1 negations are possible. Then in two-valued logic not only one but three negations are possible:

1) $N^1x = 3-x$,
2) $N^2x = 2$,
3) $N^3x = 1$.

Above we have seen various types of negations, adopted in different logical systems.

To give a survey of all possible negations understood in such a way is part of the problem of completeness of a construction: if it is functionally complete, then it also includes all functions of one argument. There is an example where this is obtained exclusively on the basis of functions of one argument. For example, all functions of one variable can be defined by means of the following three (if $n = 0, 1, \ldots, k-1$):

1) $F^1x = (x-1) \ (mod\ k)$;

2) $F^2x = x$, if $0 \leqslant x \leqslant k{-}3$; $F^2x = k{-}1$, if $x = k{-}2$; $F^2x = k{-}2$, if $x = k{-}1$;

3) $F^3x = 1$, if $x = 0$; $F^3x = x$, if $x \neq 0$ (it is assumed that $k \geqslant 3$) (cf. JABLONSKIJ).

In order that a function be called a negation of x it is normally not only required that it be a one-place function of x, but there are still further requirements. Let us first consider them in the case of two-valued logic. Here N^2x and N^3x are respectively always false and always true propositions. And, in the end, $N^2x = N^1AxN^1x = N^1N^1KxN^1x$, $(N^3x = N^1KxN^1x = N^1N^1AxN^1x$. Therefore only N^1 (we will write it simply N) is taken as negation. It assumes the following functions, which because of the two-valuedness are here essentially connected:

1) it appears in the formulations of laws like $Kxy = NANxNy$, $NNx = x$, $Cxy = CNyNx$, etc.; i.e., it is used in the expression of one function by means of others (x can be considered as an identity function);

2) it expresses that Nx is asserted if and only if x has the value 'false' (in general if x has a value corresponding to falsity);

3) it transforms an asserted proposition into a non-asserted (a denied) one.

In n-valued logic these functions become differentiated. Leaving to N the rôle mentioned under point (1), it can be defined, in particular, as $Nx = n{-}x + 1$. In this form N allows generalizations of two-valued principles (the operators C and N are sufficient for the generalization of all functions of two-valued logic). For the description of the second function a new operator has to be introduced, say J^kx where $1 \leqslant k \leqslant n$, such that J^kx is asserted if and only if x has the value k. Thus there are several functions $J^{s+1}x$, $J^{+2s}x, \ldots, J^nx$ playing in n-valued logic the rôle indicated. The cause of this is clear: in n-valued logic not only one, but several undesignated values are possible. For instance, if $n = 3$ and $s = 1$, then J^2x will be asserted, if $x = 2$. Obviously, $J^1x, \ldots, J^sx$ play the rôle which in two-valued logic is played by the identity function of x: J^1x is asserted, if $x = 1$, i.e., if x is asserted, etc. In this way $J^1xAJ^2xA \ldots AJ^nx$, or in another notation $AA \ldots AJ^1xJ^2x \ldots J^nx$, where A is written $n{-}1$ times, is a generalization of the two-valued law $AxNx$.

For the rôle of negation mentioned under the third point, a new operator must be introduced, say Lx (in ROSSER and TURQUETTE: $\bar{x}$), defined as $Lx = J^{s+1}xAJ^{s+2}xA \ldots AJ^nx$, or in another form $Lx =$

$AA\ldots AJ^{s+1}xJ^{s+2}x\ldots J^{n}x$, where A is written $n-s+1$ times. In words: x is denied if either $J^{s+1}x$ or $J^{s+2}x$ or... or $J^{n}x$ is asserted.

The proposition $AxLx$ will, of course, always be asserted as in two-valued logic (it expresses the two-valued principle 'x is either denied or asserted'). Defining with the help of L the implication $C^{**}xy$ by the equation $C^{**}xy = ALxy$, we get a function for which *modus ponens* will be valid in the general form: if $C^{**}xy$ and x are both asserted, then y is asserted also. The operator C of Łukasiewicz-Tarski does not generally satisfy this demand. Let $n>2$ and $s>1$, for example. Then for $x = s$ and $y = s+1$ we get $Cxy = max(1, y-x+1) = max(1, s+1-s+1) = 2$, i.e., Cxy and x will be asserted but y will not be asserted.

Although the functions of negation under consideration become differentiated, their connection can nevertheless be stated at least in single cases. Thus, the negation N defined by the equation $Nx = n-x+1$ assumes in some cases also the functions of J. Thus, for $n = 4$ and $s = 2$ there will be the following relation: if x is asserted (denied), then Nx is not-asserted (asserted). In general form, if the following conditions are fulfilled:

1) $n-x+1>s$ and $x \leqslant s$,
2) $n-x+1 \leqslant s$ and $x>s$,

then N assumes also the functions of J. There are also other types of negations possible which are different from N, where the connection takes other forms and has a more general character.

Let us take the negation M, defined as follows: if $1 \leqslant x \leqslant s$, then $s+1 \leqslant Mx \leqslant n$; and if $s+1 \leqslant x \leqslant n$, then $1 \leqslant Mx \leqslant s$. It transforms asserted propositions into denied ones and denied ones into asserted ones. The analogy with two-valued negation seems complete. However, $MMx = x$ does not always hold for M. For example, let 1 and 2 be designated values, and 3 be undesignated. Then $M1 = 3$, $M2 = 3$, $M3 = 1$; and in this case $MM2 = 1$. In order that $MMx = x$ be possible, it is necessary that n be even and $s = n/2$ (but even under these conditions it is possible that $MMx \neq x$). Two-valued logic satisfies this condition. Negation M is interesting as an example showing that a generalization is always one-sided, that it rests upon an analysis (here: the differentiation of the functions of negation) and the distinction of separate properties of the phenomenon analysed. In its turn, the generalization itself stimulates analysis. In the given case the many-valued generalizations of two-valued negation make us distinguish the functions of the latter. A similar influence on the conception of two-valued 'and', 'or', 'if... then', etc., emanates from their many-valued generalization.

§ 7. As has been shown, the constructions by means of many-valued matrices are to be distinguished according to the choice of basic functions and according to their definition of tautology (always asserted proposition). On these bases they can also be compared. We shall now make some remarks concerning this comparison.

If all basic functions of a system S^1 are definable by means of the basic functions of a system S^2, then system S^1 is contained in system S^2. For example, a system S^1 with the basic functions $Axy = min(x, y)$ and $Nx = n-x + 1$ (truth values $1, 2, \ldots, n$) is contained in the system S^2 with the basic functions $Kxy = max(x, y)$ and $Nx = n-x + 1$, because Axy can be defined by means of Kxy and Nx, and Nx is already S^2. The definition runs: $Axy = NKNxNy$. The comparison made presupposes that S^1, and S^2 are based on the same assumptions concerning the number of truth values and their mutual relations. If S^1 is contained in S^2, and S^2 in S^1, then S^1 and S^2 are equivalent. In the given example S^1 and S^2 are equivalent, since Kxy is definable by means of Axy and Nx: $Kxy = NANxNy$.

If at least one of the basic functions of system S^1 is not definable by means of the basic functions of system S^2, then S^1 is not contained in S^2. The expression 'not definable' means here: there is no formula (proposition) constructed by means of the basic functions of the one system, which would be equivalent with the function of the other system. For instance, the system with the basic functions C, N and T is not contained in the system with the basic functions K, A and N, since the function T is not definable by means of only K, A and N. If S^1 is not contained in S^2, then S^2 is, of course, not complete. If S^1 is not contained in S^2 and S^2 is not contained in S^1, then S^1 and S^2 are independent. Independent systems are obviously not complete.

The comparison of systems with respect to their tautologies takes into account the assumptions on the number of truth values, on the definition of tautologies (here mainly: on the splitting of the set of values into those designated or asserted and those undesignated) and on the choice of the basic functions. For instance, if S^1 is contained in S^2 and these systems have the same definitions of tautology, then the set of the tautologies of S^1 is a subset of the set of the tautologies of S^2; systems of n truth values can be considered as classes of systems with $n = 2$, $n = 3$, etc., such that correspondingly the classes of tautologies will successively include one another. Generally, the relations between many-valued systems are very diverse, and their consideration falls outside the scope of this work. As to the relation of two-valued and many-valued systems, we will below treat this question especially.

We will make one remark in passing, because it is somehow connected with the question of the relations among systems and has some interest from the philosophical point of view.

The invariance of the laws of logic with respect to the different forms of notation holds also for many-valued logic. Let us consider the following example of two variants of notation:

First variant: truth values $1, \ldots, n$ $(n \geqslant 2)$; $Axy = min(x, y)$; $Cxy = max(1, 1{-}x + y)$; 1 corresponds to truth, n to falsity (in general, the smaller the truth value, the nearer is it to truth).

Second variant: truth values $1, \ldots, 0$; $Axy = max(x, y)$; $Cxy = min(1, 1{-}x + y)$; 1 corresponds to truth, 0 to falsity.

In both cases the assertion $Axy = CCxyy$ is valid. For the first variant: $min(x, y) = max(1, 1{-}max(1, 1{-}x + y) + y)$; if $x = y$, then $y = max(1, 1{-}1 + y) = y$, since $y \geqslant 1$; if $x > y$, then $y = max(1, 1{-}1 + y) = y$; if $x < y$, then $x = max\,(1, 1{-}1 + x{-}y + y) = x$. For the second variant: $max(x, y) = min(1, 1{-}min(1, 1{-}x + y) + y)$; if $x = y$, then $y = min(1, 1{-}1 + y) = y$, since $y \leqslant 1$; if $x > y$, then $x = min\,(1, 1{-}1 + x{-}y + y) = x$, since $x \leqslant 1$; if $x < y$, then $y = min\,(1, 1{-}1 + y) = y$.

It should be noted, however, that in speaking of the invariance of logical laws, an isomorphism is presupposed with respect to the symbols by which the laws are formulated and whose definitions are taken into account in the proofs. In the example given there is a one-to-one correlation between the sets $1, \ldots, n$ and $1, \ldots, 0$ $(1{-}1, \ldots, n{-}0)$, and also between the definitions of A and C. If this condition is not fulfilled, then no invariance of the laws of logic can be spoken of. Let us take, for example, two such variants:

1) truth values $1, 2, 3$; A and C defined as above in the first variant; $Nx = 4{-}x$; 1 and 2 are designated values (propositions which have always either value 1 or 2 are accepted as laws), 3 is undesignated;

2) truth values *1*, $\frac{1}{2}$, *0*; *A* and *C* defined as above in the second variant; $Nx = 1-x$; *1* is the designated value, $\frac{1}{2}$ and *0* are undesignated.

The definitions of *N* in the two cases correspond to one another, if we have the correlation *1-1*, *2-*$\frac{1}{2}$, *3-0*. But because there is no correlation of the designated (and consequently also of the undesignated) values, we have here two different logical systems, and not only two different forms of notation of the same system. Although, for example, in the first system *AxNx* is a law, in the second it is not.

§ 8. The construction of functional systems is relatively easy, once their possibility for *n* truth values has been shown. But axiomatic constructions have the advantage of giving a method of surveying systematically all asserted (accepted, provable, etc.) propositions (or formulas). Therefore the natural tendency is to axiomatize the functional systems.

As a matter of fact, in many-valued logic the axiomatic constructions in one way or another are always based on functional ones; once they are intended as a logic of propositions and not simply as a collection of given symbols and of formation rules for new symbols. This is expressed in the following:

1) the axioms are selected from the number of asserted propositions of the functional construction; and also the rules of derivation of new asserted propositions are chosen in a way suggested by the functional construction; the axioms are verified by some matrices or by forms equivalent to them; the foundation of an axiom system can be established by giving an interpretation by matrices or in general, an interpretation which satisfies some construction by matrices, etc.;

2) the properties themselves of axiom systems are considered in relation to functional constructions, as relative ones.

Examples of the first point have been given above. We will give two further ones, and refer back to them later. In Słupecki 1946 the following axiom system for three-valued logic was formulated:

1) *CxCyx*,
2) *CCxyCCyzCxz*,
3) *CCCxNxxx*,
4) *CCNyNxCxy*,
5) *CTxNTx*,
6) *CNTxTx*.

As we see, this axiom system differs from Wajsberg's in that the latter is here completed by the addition of axioms (5) and (6) containing the operator *T*. All operators occurring in the axioms must, of course be defined. And if we are concerned with the axiomatization of a calculus of propositions, these definitions will have a functional character: *Cxy*, *Nx* and *Tx* must be considered as functions of *x* (Słupecki actually defines them this way). Moreover, in the proofs of the consistency and independence of the axiom system, for example, Słupecki gives a construction by truth tables. Similarly, Rosser and Turquette, in their system first give preliminary definitions of some propositional functions in term of basic functions (in particular of F^1 and F^2). We will not reproduce here the axiom system (more exactly: the system of axiom schemata) of Rosser and Turquette, because this would necessitate rather extensive explanations.

As already has been said: the system of Słupecki with the three basic functions *C*, *N* and *T* is sufficient for the definition of all functions of three-valued propositional logic with a finite number of arguments. And the axiom system of Słupecki-Wajsberg is complete in the following sense: every proposition formulated by means of the operators *C*, *N* and *T* and propositional variables is either a consequence of the axioms or, if added to the axioms, leads to a contradiction (we are not concerned here with the different concepts of completeness – the equivalence of all interpretations of the system, the derivability of all tautologies, etc. – as this falls outside the scope of our subject).

As an illustration of the second point, we give some definitions (this can only be an illustration and not an exposition of all possible definitions of properties of formal constructions; in particular, there are different definitions of completeness, consistency, etc., but we limit ourselves to one of the possible definitions). One construction is as strong as another, if all asserted propositions of the latter are also asserted propositions of the former. If an axiomatic construction is as strong as a truth functional one, then it is deductively complete with respect to the latter. If a truth functional construction is as strong as an axiomatic one, then the latter is acceptable with respect to the former. If an axiomatic construction is deductively complete and acceptable with respect to a truth functional one, then the two are equivalent, i.e., they define one and the same set of asserted propositions. As we have seen, there are axiomatic constructions which have no equivalent truth functional one (intuitionistic logic, systems of strict implication), and others which have equivalent truth functional ones (e.g. the axiomatic construction of Rosser and Turquette). For every truth functional construction, there exists at least one equivalent axiomatic construction (cf. ROSSER and TURQUETTE), but the converse does not hold.

The fact of the relativity of axiomatic constructions to truth functional ones is not accidental. In two-valued logic this fact remains concealed or, on the contrary, seems unessential and obvious because the two-valued character of the calculus is tacitly presupposed as something which goes without saying. But also there, when defining the symbols of functions, asserted formulas, etc., and when proving the consistency, independence, completeness (or incompleteness), etc. of axiom systems, one has recourse to interpretations, to adequate truth functional constructions. Moreover, from the point of view of the relation between axiomatic and truth functional constructions there is only one single truth functional construction (the variants are on this level unessential, cf. ROSSER and TURQUETTE), which again creates the appearance of the absolute independence of the axiomatic constructions. In many-valued logic this illusion of absoluteness is dispelled, if only because of the numerous truth functional constructions. Moreover, there are several further considerations supporting the thesis under consideration.

In axiomatic constructions the many-valued character of the calculus is not immediately visible. For example, one cannot yet say of the axiom system of Tarski-Bernays whether it is two-valued or not; in RASIOWA it proved to be four-valued. Besides that, once the propositions are divided into asserted and denied (acceptable and unacceptable, provable and unprovable), then all these constructions without exception seem to be two-valued. If in a given calculus it can neither be shown relative to a proposition that it is asserted nor that it is denied (if the proposition is undecidable with respect to the construction given), then this is simply taken as a characteristic of the calculus which does not presuppose the many-valuedness of the propositions themselves.

This circumstance shows that many-valued logic is only possible on a functional but not on an axiomatic foundation. However, this is not an argument against many-valued logic. The fact that in the axiomatic construction of some calculus (intuitionistic logic) the law of the excluded middle does not occur, makes it a calculus which needs an interpretation other than a two-valued one; similarly, in the case where operators having no analog in two-valued logic (e.g. the operator T of Słupecki) occur in the axioms. Moreover, the occurrence of an indication in the axiom system itself as to the possibility of three and more truth values repudiates the point of view criticized. Thus in the axiom $\Gamma^{n}_{1}\{C[J^{i}(x)]y\}y$ of the system of Rosser and Turquette the possibility is indicated that $n>2$ (Γ is a 'chain symbol' whose meaning is explained by the example $\Gamma^{4}_{3}x^{3}y = Cx^{4}Cx^{3}y$).

The conviction that an axiomatic construction of many-valued logic is impossible is based on a confusion between the logic to be constructed and the tools of its construction (the metalanguage of the logic given), i.e., between the fact that for the construction of a n-valued logic two-valued logic is sufficient, and the fact that the logic to be constructed is n-valued (that it admits n truth values). But even the axiom system of ordinary two-valued logic can be interpreted as the axiom system of an n-valued logic (cf., e.g., the proof of independence of the axiom system in RASIOWA).

Thus the possibility of interpreting as two-valued those axiom systems which are intended as many-valued, and interpreting as many-valued those which are regarded as two-valued, is a sufficiently convincing argument for the existence of an important connection between the types of construction considered.

It should be noted, however, that the possibility just mentioned is not unlimited. The axiom system of two-valued logic, based, e.g., on the operators C and N, will not be deductively complete if it is interpreted as an axiom system of three- or more valued logic (i.e., not all interpretations of this axiom system are equivalent). Therefore, there are only relative criteria of distinction here. To set up some single absolute criterion which would decide for axiom systems taken by themselves (taken out of context, without corresponding definitions and foundations) whether they are two-valued, or many-valued, is, to all appearances, impossible. Indeed, the search for such a criterion seems quite meaningless, since an axiomatic construction is only a convenient means of surveying the propositions (formulas, expressions) asserted (acceptable, provable) under given conditions. The only thing to be said with respect to such an 'absolute' criterion is the following: we are dealing with two-valued (many-valued) logic, if we assign two (more than two) truth values to propositions; or in a more general form: if arguments and functions take two (three or more) values from the given set of values; or passing over to a more general ontological level: if objects can have one of two (three or more) exactly identifiable properties.

Of course, the axiomatic constructions also have an autonomous value independent of the functional constructions. But we are speaking here about many-valued logic, and about which kind some given axiomatic construction of propositional logic belongs to from this point of view, something which cannot always be determined by a simple inspection of the axiom system. And if one takes into account the fact that different functional interpretations of one and the same axiom system are possible, then one excludes the existence of general criteria for such a determination. Moreover, because the main task of axiomatic constructions is to determine the classes of derivable propositions (to give a method for distinguishing between derivable and underivable propositions), the division of the propositions according to their truth values recedes into the background. It is with respect to given axioms and rules of derivation that the propositions are divided here into two (derivable and underivable) or three groups (if there are propositions with respect to which in the given calculus it cannot be decided whether they are derivable from the axioms or not). The asserted propositions can generally be defined, without reference to truth values, as the axioms and their consequences.

Therefore, the axiomatic constructions are by themselves neither two- nor many-valued. We can say that they are constructions related to investigations in the domain of two- or many-valued logic, that they are somehow connected with the ideas of these logics being tools for the solution of determinate problems of n-valued logic, that they are constructions following an n-valued conception of logic. Naturally, distinguishing them according to the composition or form of the axioms by itself says nothing about distinguishing them from the point of view of truth values. Thus the exclusion of the *tertium non datur* from the axioms or the statements derivable from the axioms does not mean by itself a transition to many-valued logic. Only in connection with the informal considerations motivating this exclusion and with the attempts to interpret (justify) it by means of a three-valued logic is it possible to speak of an informal and a formal approach to many-valued logic and to consider the ideas of Brouwer as a source and stimulus of the development of the latter.

§ 9. We have been considering various forms of the relation between two-valued and many-valued logic. And this seems quite sufficient for eliminating any one-sided point of view with regard to this question. But we will give one more example of a many-valued logic which has no analogy at all with two-valued logic and is not a generalization of it. Let *1*, *2*, *3* and *4* be truth values. We introduce the following functions:

1) $F^1xy = 2$, if $x + y$ is an uneven number; and $F^1xy = 3$, if $x + y$ is an even number;

2) $F^2xy = 1$, if $x + y$ is an uneven number; and $F^2xy = 4$, if $x + y$ is an even number;

3) $F^3x = 5 - x$.

Such a system may be conceived abstractly for some end. However, it is impossible in this system to define the functions *Cxy*, *Kxy* and *Axy*, since in the definition of the latter all the truth values should occur in the matrix. However, by means of the functions F^1, F^2 and F^3 we can only define a small class of functions in whose column of values will occur either the same repeated values or two alternating values.

The valuable words of Rosser and Turquette quoted in the beginning of § 4 of Chapter One must be supplemented by the following statement which, perhaps, seems nowadays even more pressing: Ever since systems have been constructed in logic which are based on the principle 'the propositions have *n* truth values, where *n* is any positive integer', there have been those who questioned this principle. And whereas in the first case the doubts were based on real difficulties appearing in the analysis of human knowledge within the limits of two-valued logic, and finally led to important progress in the history of logic, nothing of the sort can be said of the second case.

According to GÜNTHER two-valued logic is the only possible one. And what is called many-valued logic is for him only the description of connections between different semantical levels of our knowledge, every level of which is ruled by two-valued logic. In particular, three-valued logic appears as the connection between three aspects of two-valued logic. For instance, the truth tables of *Kxy* for which $Kxy = min(x, y)$ is presented as the unification of three tables of two-valued conjunctions

K^1xy, K^2xy and K^3xy where x and y take respectively one of the values *1* and *2*, *2* and *3*, *1* and *3*. Similarly for Axy.

However, it is not difficult to show that only a small portion of all functions can be obtained by this method. Let us take the following function (the implication of Łukasiewicz):

1) $Cxy = 2$, if $x = 1$ and $y = 2$, and if $x = 2$ and $y = 3$;
2) $Cxy = 3$, if $x = 1$ and $y = 3$;
3) $Cxy = 1$ in all the other cases.

By separating the matrix corresponding to this function into three two-valued ones (following the idea of the author), we get

a)

x \ y	1	2
1	1	2
2	1	1

b)

x \ y	2	3
2	1	2
3	1	1

c)

x \ y	1	3
1	1	3
3	1	1

But as we see the second matrix is not two-valued: in a two-valued matrix *2* should stand in the place of *1*, and *3* should stand in the place of *2*, i.e., the matrix should look like this:

d)

x \ y	2	3
2	2	3
3	2	2

Generally, in all the cases where the function takes values not taken by the arguments of the two-valued matrix, the method under consideration does not work (it works for functions which take either the minimal or the maximal value of the arguments). Moreover, if we try to compose a three-valued matrix out of the matrices (a), (c) and (d), then we get the following nonsense: for $x = y = 2$ and $x = y = 3$ the implication takes simultaneously values *1* and *2*, i.e., $Cxy = 1 = 2$ which is clearly absurd. Finally, we might mention that even if this method is applied exclusively to Kxy and Axy, it is of no use for constructing many-valued systems. In particular, operating with only the functions Kxy and Axy and two-valued negation, it is impossible by means of them to define functions like $Fx = 2$ (for three-valued logic).

The consideration of this example, of course, does not prove that it is impossible to reduce many-valued logic to two-valued logic. But here no such proof in the strict sense is needed. Many facts of different kinds are available which are enough to show that many-valued logic is not reducible to two-valued logic from some point of view or other (e.g., the fact that there are functions having no analogs in two-valued logic, that different classes of tautologies can be distinguished, etc.). Another question is the one of the interpretation of logical systems as theories of thinking. In our opinion, one may here make use of many-valued systems, but one may also dispense with any *n*-valued interpretation (the two-valued one included) (cf. ZINOV'EV 1961a). We will come back to this question in considering the interpretation of many-valued constructions (actually, we will treat it only briefly, since it goes beyond the scope of our subject).

Let us come back once more to the case where a many-valued interpretation of two-valued logic is used for the foundation of an axiom system. For instance, the independence of the axiom system of Tarski-Bernays

1) *CCxyCCyzCxz*,
2) *CyCxCxy*,
3) *CCCxyCCxyxx*,

can be proved in the following way (cf. RASIOWA):

1) For the proof of the independence of the first axiom from the remaining ones we construct the matrix

C	0	1	2
0	1	1	2
1	0	1	2
2	0	2	1

The second and the third axiom satisfy it (they always take the value *1*), but not the first, since in the case $x = 0$, $y = 2$ and $z = 0$, the first axiom takes the value *2*.

2) For the proof of the independence of the second axiom from the remaining ones we construct the matrix

C	0	1	2	3
0	1	1	2	2
1	0	1	2	2
2	0	1	1	1
3	0	0	0	1

The first and third axiom satisfy it, but the second axiom takes, in the case $x = 1$ and $y = 3$, the value 0.

3) The independence of the third axiom is verified by the matrix

C	0	1	2
0	1	1	1
1	0	1	2
2	0	1	1

which the first and second axiom satisfy, whereas the third axiom takes, in the case $x = 2$ and $y = 0$, the value 2.

These interpretations may be considered as uses of the ideas of many-valued logic for deciding problems of two-valued logic.

§ 10. But independently of the relations considered the following two general questions arise:

1) Can an n-valued system be constructed in which the negations of the laws of two-valued logic would be assumed or could be derived, particularly the negations of the law of the excluded middle and of the law of contradiction?

2) Is two-valued logic not only sufficient but even necessary for the construction of an n-valued system?

The answer to these questions is necessarily connected with the answer to the question: in what sense are the laws of two-valued logic preserved in n-valued logics and in their construction, and in what sense are they not preserved?

The answer to the first question can be given within the limits of logic itself. If a n-valued logic is constructed as a generalization of two-valued logic (such that for $n = 2$ we get two-valued logic), then it goes without saying that the negations of the laws mentioned will have values corresponding to falsity, if x has a value corresponding to truth. However, independently of these considerations, the negations of these laws will not be laws of n-valued logic.

We define the operators A and K, by means of which the laws under consideration are constructed, as follows:

1) Axy is asserted, if at least one of x and y is asserted; and it is unasserted, if x and y are both unasserted.

2) Kxy is asserted if and only if both x and y are asserted.

Let us take $NAxNx$ where Na is any one-place function of a. The idea is that this proposition should be asserted for any value of x. With this assumption two cases are possible:

1) N transforms a from an unasserted into an asserted proposition.

2) N leaves a asserted.

In the first case it is obvious that $AxNx$ should be unasserted. And this means both x and Nx must be unasserted. Then $Na = j$, where $s \leqslant j \leqslant n$, i.e., Na has always an undesignated value (is always unasserted). That means N does not transform a into an asserted proposition, which contradicts the assumption; or we can also obtain a contradiction following another line: because of this property of N, $NAxNx = j$, i.e., it is unasserted, which contradicts the assumption. In the second case $AxNx$ should also be asserted. But here $Na = k$, where $1 \leqslant k \leqslant s$, i.e., the negation does not fulfil functions similar to those in two-valued logic. Here N is not a negation in the proper sense of the word. Similar arguments can be set up concerning the second law.

But what has been said will suffice. And when, moreover, we remember that consistency is demanded of many-valued constructions (and consistency is defined, in particular, as not admitting simultaneously the negation of x and of Fx) then the picture becomes sufficiently complete. Reasoning informally: to accept the negations of the laws under consideration as asserted is equivalent to permitting simultaneously in logic an assertion x and its negation.

In ROSSER and TURQUETTE, for instance, the consistency of a construction is thus defined: the construction is F-consistent, if there is no x for which x and Fx are both denied, where Fx is some propositional function constructed by means of basic F_i; a construction is consistent, if there exists a proposition which is denied according to the given construction. These definitions are generalizations of corresponding two-valued ones. If the negations of the laws of two-valued logic under consideration are now admitted, then in such a construction there will be no denied propositions, and together with this, x and Fx will be negated simultaneously, which leads to an incredible confusion and makes it impossible to give an interpretation of the calculus for any domain of objects whatsoever.

Therefore, the relative character of the laws of logic must not be understood in the vulgar sense, i.e. as the possibility of their negation in any cases whatsoever.

§ 11. The second question raised above takes us outside the internal problems of logic.

Whatever may be the terms by which we construct n-valued logical systems (either terms of mathematics, traditional logic, or ontology), ordinary language is necessary and two-valued logic is sufficient for their construction. As for ordinary language, its rôle is obvious. And that two-valued logic is sufficient is shown by the facts and by experience in constructing n-valued systems. However, there is here a 'but' which we should first of all elucidate.

In saying that two-valued logic is sufficient for the construction of n-valued logic, we mean first of all that any function of n-valued logic can be described with the help of propositions having a two-valued character (cf. also ROSSER and TURQUETTE, ŠESTAKOV 1946). Let us take, for example, the function Fxy given by the table

x \ y	1	2	3
1	2	1	2
2	1	1	3
3	3	3	3

The very notation shows here that if a proposition *has* the value i, then it is not the case that it does *not* have this value (writing i in some place of the matrix we do not cancel it at this same place but fix it there). Instead of giving the table we can describe this function by equations:

1) $Fxy = 1$, if either $x = 1$ and $y = 2$, or $x = 2$ and $y = 1$, or $x = 2$ and $y = 2$;

2) $Fxy = 2$, if either $x = 1$ and $y = 1$, or $x = 1$ and $y = 3$;

3) $Fxy = 3$, in all the other cases.

Thus all propositions of the form $a = i$ are two-valued: either what they say is in fact the case or a is equal to another truth value, different from i. Finally, symbolizing by ai the fact that proposition a has the value i, we can write this function in the form (for the sake of clarity we place the symbols of functions *between* the propositions):

$\{[(x1Ky2)B(x2Ky1)B(x2Ky2)]\ C(Fxy)1\}K$

$\{[(x1Ky1)B(x1Ky3)]\ C(Fxy)2\}K$

$\{[(x2Ky3)B(x3Ky1)B(x3Ky2)B(x3Ky3)]\ C(Fxy)3\}$, where K, B and C are respectively two-valued conjunction, strong disjunction and implication. Because *x1 is true* $= x1$, *x1 is false* $= x2Bx3$, $Nx2KNx3 = x1$, etc., we can reconstruct the truth table by looking over all possible combinations of truth values of x and y (strong disjunction, we remember, is true if and only if exactly one of its members is true) and by using the rules:

1) if Kab is true, then a (and b) is true;

2) if Cab and a are true, then b is true.

For example, let $x = 1$ and $y = 1$; from the given full notation of the function, it follows that $[(x1Ky1)B(x1Ky3)]\ C(Fxy)2$; since for $y1$ it will be false that $y3$, therefore $x1Ky3$ will also be false; since $x1Ky1$ will be true, the disjunction will be true and $Fxy = 2$ will be true. What has been said here is shown in ŠESTAKOV 1946 and ROSSER and TURQUETTE in a slightly different form (in the latter the method of 'partial normal forms' does not use the exclusive disjunction B but the disjunction A).

Even for a more complicated case, where the possibility of writing simultaneously different values in one place of the table is admitted, the situation is essentially the same. For instance, a function is possible, in whose description we find such expressions as $[(x1Ky2)B(x2Ky1)]\ C\ [(Fxy)1B(Fxy)\ 2]$, or in another form, Fxy equals either 1 or 2, if either $x = 1$ and $y = 2$, or $x = 2$ and $y = 1$.

Reasoning in an abstract way, it is not difficult to show that an n-valued calculus can be constructed by using a language which is subjected to an m-valued logic where $m>2$. But since we are not accustomed to using an m-valued logic, we must first construct it. Also the solution of this new task must again be achieved with the help of some logic, etc. – as long as we are not finally giving up the hopeless attempt of exhausting the infinite and do not rely on common sense or the routine thinking we have acquired by experience, knowledge of things, education, etc. Moreover, the legitimate question arises: in what type of m-valued logic is n-valued logic to be constructed?

The construction of an n-valued logic is the solution of an informal problem. As in every scientific investigation, definitions are introduced here and assertions are accepted on whose basis the system of assertions is developed. Assertions are accepted as true if they are proved. If an error is admitted, then the corresponding judgments are counted as wrong. Errors are thus cleared up, once some true judgments are proved. Meaningless expressions are eliminated at the very beginning by the definitions. For instance, the sequence of symbols xyK is eliminated because according to the definition of 'proposition' (or of 'formula') it is not included in the number of propositions. From the fact that in an n-valued construction a class of propositions can be distinguished such that it is impossible to prove in the calculus either that these propositions are asserted or that they are denied, it does not follow that supplementary truth values are absolutely necessary. The case is similar, if this problem itself (the problem of decidability) proves to be undecidable. A purely informal characterization of the propositions and the properties of the calculi is here quite sufficient (of course, only as long as such problems do not become the object of a generalized investigation). Thus, a stronger statement than the one concerning the sufficiency of ordinary two-valued logic can be made: to construct an n-valued calculus with the help of an m-valued logic would really mean complicating the problem without justification.

In saying that two-valued logic is sufficient for the construction of n-valued logic, we mean secondly the ordinary way of operating with the rules of logic, a way which can be noted even in the investigations of scholars who never studied logic (the propositional calculus included), and which also proves necessary for the construction of logic itself (the classical calculus of propositions included). Obviously, two-valued logic is meant in another sense here (cf. GRENIEWSKI 1957). We will consider this in more detail.

§ 12. The question raised in the preceding section goes beyond the scope of our subject – it is a philosophical question concerning the nature and construction of logic in general. Moreover, it is too many-sided to be systematically comprehended here. Therefore, we limit ourselves to fragmentary remarks which have a direct bearing on our subject.

The well-known 'paradox' of logic (that for the construction of logical theory it is necessary to make use of logical theory) proves troublesome only in the case where deductive method is the only method admitted for constructing logic. However, in the history of science this difficulty is in fact minimal. Indeed, in the construction of different sciences in the majority of cases no reference to logic is made. One simply relies on the acquired routines of thinking which finally impress people as something objective, as something conditioned by the very properties of the things reflected and of the organs and means of thought reflecting the things. This shows that, explicitly or implicitly, logic is based on some empirical considerations and observations. The aspiration to formulate these in a theoretical system constitutes an essential part of traditional logic, understood as a philosophical discipline.

The disdain of contemporary logic for this sort of question is connected with different kinds of erroneous philosophical positions, like, e.g., the ones which defend the *a priori* character of classical logic (that it preceeds experience) or the arbitrariness in the choice of a logic. The development of many-valued logic gave a heavy blow to apriorism in logic (cf. MOSTOWSKI), similar to the one struck by the geometry of Lobačevskij in mathematics, because it showed the possibility of logical systems which are different from classical logic. With respect to the position defending the arbitrariness in the choice of a logic, a criticism consists in the analysis of the facts concerning the interpretation of logical theories and in the consideration of the question brought up above.

The disregard for the empirical character of the foundation of logic is justifiable only to the extent to which the theories developed in contemporary logic (developed under conditions where necessarily the empirical foundations of logic explicitly or implicitly played their rôle) allow a strict formulation or an explication of the intuitive considerations resting on this empirical basis. Thus, for example, the laws of identity, contradiction and the excluded middle of two-valued propositional calculus are an explication of the corresponding laws of traditional logic.

From a purely formal point of view one can easily get into a vicious circle here, because the definition of the truth functions with whose help these laws are formulated, depends itself on acknowledging the two-valued character of propositions. However, from the point of view of the development of science, there is no vicious circle. An examination of the lower levels of science by means of tools of the higher levels is quite normal and gives useful results, as it rests on the knowledge and routines of thinking already gathered.

But every explication is one-sided, since it makes use of a formal model of the material explicated (cf. ZINOV'EV and REVZIN). Thus, in the case of the laws of propositional logic, only the determinate connection between the propositions is abstracted, whereas the relation of the propositions to reality is left out: truth and falsity are accepted as undefined (their definition is made difficult, among other things, by the fact that the internal structure of the atomic propositions is disregarded). Thus, a greater generality than in traditional logic is achieved: the truth and falsity of any form of propositions are now considered, whereas in traditional logic the conception had been centered around a definite propositional structure, the attributive one (attributivity being understood not as a property of a limited class of propositions, but as a property of propositions in general; cf. ZINOV'EV 1959c), and even that structure had been taken in a very simple form (applicable to propositions concerning the presence or absence of a single property in an object). But we repeat that this is obtained by disregarding the intuitive or descriptive understanding of the truth values themselves of traditional logic.

Every explication presupposes further that the material explicated is given and can somehow be described. What is the character of logic on the empirical level, so to speak, for which formal models are introduced and an explication is given? In answering this question we must take into consideration the real facts and not abstract, hypothetical possibilities.

Among the general properties of things and of the means presenting them are some which lie at the basis of the laws of logic as understood in the above sense. In particular, the following facts exclude complete arbitrariness in the choice of a logic: it is impossible that at one and the same time an object having some property should not have it (or an object not having some property should have this property); an object either has a given property or does not have it, etc. One thing is that we exclude from among the meaningful propositions such propositions as 'an even number is not even', which shows already the working of principles which reflect such facts of everyday experience.

The point is not that we are concerned here with absolute laws: objectivity is not absoluteness. The point is that we are concerned here with very simple generalizations which in one way or another are forced on man in his everyday activity. The investigation of more complex cases where the need for another type of generalizations is revealed, is based on the simpler generalizations. Such more complex cases are, for example, the cases where it is difficult (impossible or inexpedient) to establish the presence or absence of properties in the given objects because one is considering very transitory or border-line states of them (this example is treated in detail in ROSSER and TURQUETTE), or where one allows for more than two possible states of the objects, etc.

Thus, one must distinguish between a purely negative attitude toward many-valued logic which tends under one or another pretext to reject obvious facts, and an attitude toward it as a legitimate stage (or better: fact) in the development of thinking and a part of the system of the science of logic as a whole.

If a correspondence between properties and sets is established (if it is accepted that the attribution of a property to an object and the inclusion of this object in the set of objects possessing this property are different forms of expressing the same thing), then what has been said can be represented in the following simple and convenient schema. The division of some given non-empty set of objects into two non-overlapping subsets which exhaust it, is a very simple case in the examination of the relations between sets. This principle of division remains valid in all other cases. Not in the sense that all relations between sets are of this kind, but in the sense that any of their relations can be represented as a combination of such two-membered divisions.

For example, let the set of the objects A be singled out (be given) and let us examine the relations of sets with respect to this set. First of all, by saying that some set of objects A is given, we are thus dividing the set of objects in the following sense: to select anything whatsoever means to distinguish it from everything else. Thus we either include some object in this set or we do not, (we either attribute some property to the object or we do not). Further, let us take a division of A into two intersecting subsets B and C. We get three subsets of objects of A, namely

1) the set of objects belonging only to B,
2) the set of objects belonging only to C,

3) the set of objects belonging to both B and C.

In terms of two-membered divisions this would look as follows:

a) the set A is divided into two sets of objects: the objects belonging to the set (1) and those not belonging to it;

b) similarly for (2);

c) similarly for (3).

In exactly the same way we can proceed also in all the other cases and convince ourselves that the two-valued principle is observed in the following form: an object is either included in some set A or it is not included (the latter can be considered as inclusion in the set not-A). Thus inclusion and non-inclusion are not understood as the existence or non-existence of the possibility of affirming something about the object, but they are something objective: the object either belongs to the set A or it does not belong to it (i.e., it belongs to the set not-A). It should also be noted that, if any two sets A and B are given whose relationship is unknown (and this makes no difference here) then the membership of an object in B certainly does not mean that this object does not belong to A (likewise in terms of properties, the presence of some property in an object does not necessarily mean the absence of another property).

Applied to the truth values of n-valued logic, what has been said means (in a very general sense): if $x = i$ $(i = 1, 2,\ldots n)$, then it is not true that the value of x is not equal to i. For instance, ascribing the value 1 to x in some complex expression, we do not in the same act of evaluation (simultaneously) ascribe to x a value different from 1. If we were to admit this, then the whole expression could become unrecognizable. In short, if one admits such a possibility, then all clarity of thinking disappears and thought cannot be fixed at all.

This general principle can be specified further in the following conditions:

1) if $x = i$, then $x \neq k$, where $i \neq k$;

2) if $x = i$ and $x = k$, then $x \neq j$, where $i \neq j$ and $k \neq j$;

3) if $x \neq i$, then $x = k$, where k is one of the values $1, 2,\ldots, n$ and $i \neq k$;

etc. All regulations of this type are nothing but the carrying over of the general two-valued principle to cases where at least one of the two possibilities is further differentiated. Two-valued logic did not consider this.

If any relations are established among truth values, for their part, then a similar result can be obtained again by numbering these relations. For example, if we have values *1**, *2**, ..., *n** such that *1** = *A23*, *2** = *C34*, etc., then we can construct a system on a general basis for *i**. What this system translated into *i* will be like, will depend on the relations of *i* (of course, these relations should satisfy the postulate of consistency).

Such a very general conception of two-valuedness, with respect to which *n*-valuedness itself appears as a particular case (i.e., as a case which can be described as a differentiation of two-valuedness), does not always conform to two-valuedness in the traditional sense. Let us consider some of the gradations of these concepts.

The principle of two-valuedness, as understood in the very general sense, finds different expressions in logic. First of all the propositions reflecting the objects naturally fall into two classes if we consider their relation to the objects described in them: 1) they are true, if they assign the objects to just that set to which they belong (if the actual state of affairs is as is described by the propositions); 2) they are not true or false, if they assign the objects not to that set to which they belong (if the actual state of affairs is not as is described by the propositions). Secondly, the set of propositions, simply as a particular case of a set, is divided into two subsets one of which is called the true propositions and the other the not-true (false) propositions. Thus, in making this division in the propositional calculus, we are not considering the definitions of these properties of the propositions (as relations between the propositions and the objects).

In both cases one of the two sets is considered as the negation of the other, and the principle of double negation is valid (negation is determinate). The states of affairs are, on this level, only taken into account as far as the objects are ascribed to the one or the other set. For example, asserting that a given contact is not in the position *1*, we describe a state of affairs. Thus, there are again different possibilities: that the negation of position *1* means the assertion of position *2* (if there are only two possible positions), or that, in the case of three or more possibilities, it means the assertion of any other position different from *1*. Generally one thing remains valid: either position *1*, or not *1* but some other.

However, the process of knowledge (from the point of view of our subject) develops so that in each or in one of the two possibilities its gradations are fixed whose calculation proves important for knowledge. Let us take such an example. Suppose some given set of objects has been divided, according to the principle of two-valuedness, into *A* and not-*A* (we assume, by the way, that the words 'given set' correspond to a state of affairs where the object domain under investigation is in some way distinguished or delimited). Suppose not-*A* to be divided in the same way into *B* and not-*B*. Suppose further that the proposition *x* ascribes some object to *A*. We take the ordinary two-valued negation of *x* (we discover that the object is not an element of *A* and does not have the property attributed to it). This *Nx* will describe an actual state of affairs, as it ascribes the object to not-*A*. However, if this does not seem sufficient and one wishes to determine whether the object belongs to *B* or to not-*B*, then without further investigation the two-valued negation cannot give a description of the case: with respect to *B* and not-*B* it is not determinate in the sense that *Nx* is either *y* or *Ny* (where *y* is the proposition ascribing the object to *B* and 'either ... or' is strong disjunction). Similarly *Ny* will either be *x* or *y*. And similarly for the cases where *A* is divided into *C* and not-*C*, where *B*, not-*B*, *C* and not-*C* are divided, etc.

The same schema is followed in accounting for the complex structure of propositions. Two-valued logic basically concentrated on the simplest attributive propositions 'object *a* has the property *P*' and 'object *a* does not have the property *P*'. This articulation is completely isomorphic to the division of a set into two complementary subsets *P* and not-*P*, since the principle of double negation is preserved. But let us take the proposition *Kxy* and its negation meaning that the state of affairs is not as it is described by *Kxy*. Evidently, the set of possible states of affairs must then be divided as follows:

1) the set of states of affairs as described by *Kxy*;

2) the set of states of affairs as described by *NKxy*; where the latter set can again be divided into

a) the set of states of affairs as described by *KNxy*;

b) the set of states of affairs as described by *KxNy*;

c) the set of states of affairs as described by *KNxNy*.

In all such cases an additional definition of negation applicable to the structure of the propositions is needed which does not rely on common sense.

In traditional logic similar questions have been treated in connection with the establishment of different types of negation which are not equivalent with one another. Let us take, for example, the judgment. 'John is reading a book'. We deny this judgment by saying that it is not true. What does this negation mean? Obviously there are several possibilities:

1) Somebody is reading, but it is not John.
2) John is not reading at all.
3) John is reading, however he is not reading a book but, say, a journal.

When asking for logical emphasis, then in fact the conditions of only one determinate negation were given. As a matter of fact one thing remains valid in all the cases: if the original proposition is true, then all the possible negations of it are not-true.

From what has been said it appears that within the limits of classical logic we have a twofold conception of two-valuedness and negation:

1) the negation is understood as indeterminate (as we will say), i.e., Nx means: not as is said in x, and the principle of the excluded middle has the form: either proposition x has the truth value i $(x = i)$, or it does not have it $(x \neq i)$;

2) the negation is understood as determinate, i.e., the negation asserts the second out of two precisely defined states of affairs, and the principle of the excluded middle has the form: either $x = i$ or $x = k$.

Normally two-valuedness is associated with the second conception, although in traditional logic these things were not strictly distinguished. Two-valuedness in the second sense is a particular case of two-valuedness in the first sense (it is even more particular if it is formulated only in terms of 'true' and 'false'). Two-valuedness in the first sense remains completely valid also in the constructions of n-valued logics. Moreover, it is indispensable as a natural condition of all scientific thinking.

It is reasonable to ask, why does one not begin with one-valued logic, with a logic without negation? In a certain sense this is in fact done: the definition of such concepts as correspondence, truth, etc., does not presuppose two-valuedness (falsity itself can be defined by means of truth). In any case, the correspondence between knowledge and reality, the empirical basis of logic, is as evident as the subsequent account of facts of non-correspondence.

§ 13. The explanation of the mutual relations between two-valued and many-valued logic, of the mutual relations between different many-valued systems themselves, of the mutual relations between different ways of constructing logical systems – all this together constitutes the solution of the general question of the multiplicity and the unity of logic.

The multiplicity of logic is a fact expressed by the existence of a great number of different logical systems – a fact which was not familiar to traditional (pre-mathematical) logic, if one does not take into consideration the differences in the method of exposition or in the completeness of comprehension which may occur in the treatment of one and the same material. It was the rise of many-valued logic which made this multiplicity more deeply felt. Naturally, there arises the problem of the unity of logic as a science.

The unity of logic, like that of any science, cannot be established on the basis of any purely *a priori* postulates or forever. It can only be established on the basis of concrete investigations of logical theories as they occur in this or that stage of the development of logic. In order to draw some conclusions from what has been said above, some general remarks on this account will now be made.

First of all it follows that the unity of logic is based on observing (as a standard, of course) the postulate of consistency not only inside its particular theories but also in comparing different theories. Here the situation is as follows. Suppose x is an axiom in one theory but is not an axiom in another. This circumstance signifies that the logical theories have a different content, that they are differently oriented (if they are not equivalent), etc., but it by no way suggests the existence of a logical contradiction. The latter case arises only if in one logical theory x is an axiom, and in the other, Nx, the negation of x such that the conjunction of x and Nx forms a contradiction (has no interpretation, does not represent anything existing). Such a situation would have to be admitted, if the admission of the legitimacy of logical contradiction were useful as a reflection of something real: the axiom systems of different theories can be united to get a new system which is intended to have some interpretation in some domain of objects. Moreover, no one is forbidden to construct theories which contain among their axioms the negations of axioms of theories which are practically and theoretically confirmed or admit the possibility of rational interpretations. But these theories are known to be empty. Logic excludes such theories on the same grounds on which every science excludes phantastical fictions which in principle lack any real content.

Above we have seen that what has been said applies to the mutual relations between two-valued and many-valued logic as well as to the mutual relations between logical theories occurring inside many-valued logic. Moreover, only on the basis of observing the postulate of consistency (which, we have seen, can be generalized in the many-valued sense) is it possible to establish between logical theories such relations as the relation of generalization (e.g. in ROSSER and TURQUETTE the work of Słupecki is generalized), of inclusion (e.g., the logic of Bočvar is included in the logic of Kleene), of equivalence (cf. the different axiom systems of intuitionist logic), of isomorphism (e.g. the axiom system of two-valued logic and some system of consequences of the logic of Bočvar), etc. The solution of one of the aspects of the problem of the unity of logic consists especially in the explanation of such relations.

The unity of logic finds further expression in the fact that the logical theories themselves become the subject of generalizing investigations. The principles of their construction, of their differences and relations, the general properties and the particularities of separate groups of them, etc. are studied (as done, for example, in the works of Rosser and Turquette, Słupecki, Jablonskij, Suszko and many others). Thus two-valued logical systems are considered as particular cases of *n*-valued ones.

Lastly, the unity of logic is expressed by the fact that there are some general concepts and principles which lie at the basis of two-valued and many-valued logic, at the basis of various logical systems. We by no means have in mind some stable and generally recognized condition of the science of logic, but rather a necessary tendency in the development of logic, which does not exclude a variety of points of view on logic, on its methods of construction, on the breadth of the material covered, etc. It is only that in very different logical systems we deal with the concepts of proposition, truth, falsity, truth value in general, derivation, etc. (or correspondingly with formulas, arguments, functions, values, transformations, etc.), which shows that the logical theories are unified in a kind of a whole, namely in the science of the structure of knowledge, of derived knowledge, of demonstration, etc.

In logic the question has been sufficiently clarified concerning the impossibility of reducing such a unification to a unification within the limits of a single axiom system (cf. the works of Gödel, Church, Rosser, *et al.*). It is sufficient, for example, to take the axiom system of two-valued logic and the axiom system of some many-valued logic which are not equivalent: their unification would, of course, bring about the loss of the specific aim of each, not to speak of the fact that within the limits of a single axiomatic construction it is impossible (by its proper means) to investigate the properties of the logical theories themselves (their consistency, independence, etc.). The unity of logic on the level considered is achieved by very different means, about which we have spoken above more than once; actually it is achieved in the way in which the unity of different theories and parts of science is achieved on the basis of the unity of the object domain investigated.

CHAPTER FOUR

TWO-VALUED AND MANY-VALUED LOGICS

§ 1. What has been said in the previous chapter on the mutual relation between two-valued and many-valued logics can be suitably supplemented by approaching this question from a somewhat different point of view (cf. the appendix in ZINOV'EV 1960, and ZINOV'EV 1962).

We adopt the following symbolism:

1) $x, y, z, \ldots$ are propositional variables;

2) $1, 2, \ldots, n$ are truth values;

3) i is any of the values $1, 2, \ldots, n$;

4) $x = i$ means x has the truth value i; $x \neq i$ means x does not have the truth value i.

The functional construction of propositional logic (or, in the terminology of NOVIKOV, the construction of the algebra of propositions) represents a special solution of an informal problem, although the very terms 'proposition', 'truth value', etc., are taken as undefined or are even replaced by a more general terminology. It is the solution of an informal problem in the sense that it starts from certain informal hypotheses which are not always clearly formulated. Besides the hypothesis about the number of truth values (and about the acceptable way of designating them), one must here mention, first of all, the hypothesis about the mutual relations between the truth values. All known constructions start in this respect from the following presupposition: either $x = i$ or $x \neq i$; it is impossible that simultaneously $x = i$ and $x \neq i$. And in this sense they start from the presupposition of the two-valuedness of the propositions $x = i$. At the same time, they also start from the following presupposition: either $x = 1$ or $x = 2$ or ... or $x = n$. And in this sense they start from the presupposition of the n-valuedness of x.

These two presuppositions should be strictly distinguished: one of them is concerned with the means of the reasonings which we make in constructing a given system, the other characterizes this system itself. The first presupposition is also the basis for using two-valued logic in constructing n-valued logic. In constructing two-valued propositional logic the definitions and assertions of this very logic under construction, naturally, cannot be used: one must rely here on the customary habits of thinking.

In constructing n-valued logic, however, the functions defined in two-valued logic can be used; we have here already a higher level of logical science. Thus we have in mind the use of two-valued functions as tools of reasoning in describing the functions of the n-valued construction.

With respect to what has been said, several important remarks should be made. In constructing n-valued logic the two-valued logic of propositions can be used. But this is not obligatory, as is shown by the facts; moreover, in this case one can make one's reasonings on the level of the customary rules. Properly speaking, in constructing n-valued logics, two-valued propositional logic is only used in so far as it is an explication of these customary rules of reasoning. In particular, the first presupposition mentioned above can be written in the form $A(x = i)(x \neq i)$ and $NK(x = i)(x \neq i)$ where A, K and N are respectively two-valued disjunction, conjunction and negation.

The construction of n-valued logics as mentioned above, is the solution of an informal problem. And from this point of view the use of two-valued logic cannot be taken as a simple coincidence. Suppose that n-valued logic is constructed with the help of m-valued logic where $m>2$. This means that in describing the functions of n-valued logic the functions of m-valued logic are used. This presupposes the admission of the m-valuedness of the propositions $x = i$. As x is given, only i can be problematic: has x the value i or does it not have it? Besides these two possibilities, a single further possibility can be admitted: it is unknown, it is meaningless, it is undecidable, etc., whether $x = i$ or $x \neq i$. And this possibility makes no sense in the definition of the very functions of n-valued logic. Only the following statement is meaningful here: if in the definition of some function a corresponding value of the function is not given for every combination of values of the arguments (although all the combinations of values of the arguments are given), then we have the definition of a class of functions. But this has no connection with a many-valuedness (with a three-valuedness) of $x = i$.

But let us nevertheless assume that the function $F(x^1,\ldots)$ of n-valued logic describes the expression Q, which is constructed from expressions of the form $x = i$ (from propositions about the values of arguments) and from symbols of the functions of m-valued logic. What is the purpose of such a description? Obviously, it consists in fixing, by means of some reasonings, the value of $F(x^1,\ldots)$ for every combination of values of the arguments. Thus two ways are possible: 1) the two-valued way where it is assumed that '$x = i$' is true if $x = i$ and '$x = i$' is false if $x \neq i$; 2) the many-valued way where '$x = i$' $= i^*$, if $x = k$, where k is one of the values $1, \ldots, n$ ascribed to x, and where i^* is one of the values ascribed to '$x = i$'. In the second case this should be done for all of the n and for all of the m values, a rather cumbersome task. But the second way should thus be some generalization of the first, in order that the purpose of describing F be satisfied in an unambiguous way. In principle such a way is conceivable. But even if its realization is admissible, it is not expedient: for solving the task in question it has no advantage whatsoever over the first way.

If to all that has been said we add the consideration that finally we will nevertheless have to rely on the customary rules of reasoning, described first of all by two-valued logic, then the rôle of the latter in constructing n-valued systems can be characterized by a 'stronger' expression than by only the assertion of its expediency (cf. ROSSER and TURQUETTE). We are not speaking here about its necessity (that it is the only way); what is important is not a theoretical proof (we repeat that another way is conceivable or admissible), but is concerned with empirical facts, with a genetic approach to the exposition of science, with expediency, etc.

§ 2. A second group of questions connected with the problem of the mutual relation between two-valued and many-valued logic deal with the comparison of their laws. Let us consider this group by the example of the law of the excluded middle and the law of contradiction. It is no accident that these particular laws are chosen: in traditional logic they were considered as basic; the doubt concerning the universal validity of the law of the excluded middle was one of the sources for the formation of the idea of many-valued logic; the laws of the excluded middle and of contradiction are immediately connected with the hypothesis of two-valuedness, etc.

The expression 'logical law' is not unambiguous. It may refer to the presuppositions about the number of truth values and their relations which lie at the basis of the functional constructions; or to the tautologies (to the asserted, always true, etc. propositions, formulas, expressions, etc.) of the functional constructions; or to the axioms of the axiomatic constructions. These different shades of the meanings of the expression 'logical law' must necessarily be distinguished in the corresponding reasonings. If we take, for example, two-valued logic, then we have the following picture. The basic hypothesis of the functional constructions can be formulated as follows: the propositions take their truth values from either of two: true and false; every proposition is either true or false; true propositions cannot be false and false ones true (it cannot be that a proposition is simultaneously true and false); in other words, the set of all propositions is divided into two non-overlapping subsets of true propositions and false propositions. Among the tautologies of two-valued propositional logic there are such which, expressed in words, say: either x or not-x; it is not the case that x and not-x. In considering many-valued systems and their relations to two-valued ones, these differences acquire considerable importance.

The comparison of logical systems according to their basic hypotheses about the number of truth values and their mutual relations presents no problem: agreement and differences are here obvious. To repeat, the argument consists in the fact that for any number of truth values the following statement is valid: either the proposition has some truth value out of the number of possible ones, or it does not have it. This statement is not rescinded in the case where even such a relation is admitted: if $x = i$, then $x = k$ (from the fact that $x = i$ it follows that $x = k$), where i and k are different values from the number of possible ones. In other words, many-valued constructions do not conflict with such an assumption. As to the comparison of tautologies, in order that it be sensible it is necessary to establish a correspondence between the definitions of the functions. Without this one can only speak of different functions and different states of affairs. The correspondence between the logical laws (and between their interpretations) should also be established in the case of the comparison of axiomatic systems. What has been said will become clearer from further examples.

§ 3. Let us take the law of the excluded middle. It can be formulated in several different forms, at least in the following three:

1) every proposition is either true or false (in this form the law was first exactly formulated by the Stoics);

2) every proposition either has some truth value or does not have it (the law can be formulated in this form, taking into account the experience of constructing many-valued propositional logics);

3) *AxNx* (*x* or not-*x*) is an always asserted (in particular: an always true) proposition or a tautology.

It is quite clear that we have here not one and the same law, differing only in its formulation, but rather various meanings of the expression 'law of the excluded middle', different statements. And the fate of each in many-valued logic (in relation to two-valued logic) is different.

It can easily be seen, that the second formulation can be applied to truth and falsity in the following sense: every proposition is either true (false) or not-true (not-false). And only when the number of truth values is limited to two (true and false) do we get the first formulation as a particular case. Thus, the second formulation is more general than the first.

The law of the excluded middle taken in the second formulation is not rejected in many-valued logic. Without it it would be impossible to construct a logical system: if we ascribe some truth value to a proposition, then what has been accepted cannot be rejected in the very same act (e.g. in the evaluation of a complex proposition, in the verification of a tautology, etc.) without running into contradiction. In the first formulation the law proves to be relative, for in *n*-valued logic it will not be valid because there the following holds instead: a proposition either has the truth value *1*, or *2*, or..., or *n*; in general: it will have one of *n* possible truth values, where *n* is any finite or infinite number.

The law of the excluded middle in the third formulation is on the one hand more general, as it concerns any proposition and its negation and not only the propositions about the truth values of propositions, i.e., a special kind of proposition. On the other hand, this formulation of the law is based on a presupposition about the number of truth values of the propositions and the definitions of disjunction and negation. Here is not the place to discuss how this seeming paradox is solved. More important is the second point, namely the dependence of this third formulation on the basic presuppositions and definitions, from which the relative character of the law is a consequence.

In many-valued logic numerous cases are known where the disjunction *A* and the negation *N* are defined as generalizations of the corresponding functions of two-valued logic, but where *AxNx* is not a law. This holds, for example, for the three-valued logic of Łukasiewicz and its generalization to any number of truth values (cf. ROSSER and

TURQUETTE), and others. The functions of many-valued logic should necessarily be generalizations of the two-valued functions mentioned, otherwise the basis for their comparison is lacking. In other words, the following condition should be fulfilled. Suppose the truth value *1* corresponds to truth in two-valued logic and *n* to falsity. Independently of the number of truth values in some construction the following should hold for *A* and *N*: $N1 = n$, $Nn = 1$, $A1n = 1$ (if a tautology has the value *1*). This condition, however, is only a condition for the comparison of the assertions of various systems. In order that *AxNx* be a tautology the following is necessary in many many-valued systems (if $1, \ldots, s$ are the designated values): if $s+1 \leqslant x \leqslant n$, then $1 \leqslant Nx \leqslant s$; $Axy = min(x, y)$.
It should be noted that the third formulation of the law of the excluded middle is not necessarily invalid in many-valued logic. This depends on the character of the generalization of the two-valued functions (here great variety is possible) and on the definition of tautologies. In the four-valued logic of Łukasiewicz, for example, the third formulation of the law considered remains valid, although the first formulation does not: a proposition has here either the truth value *1*, or *2*, or *3*, or *0*. As we see, the situation is such that it cannot be analyzed without strictly distinguishing the various meanings of the expression 'law of the excluded middle'. Even within the limits of one and the same logical system, different generalizations of the two-valued functions *A* and *N* are possible, such that on one level the generalization of *AxNx* is not a law and on the other it is a law (e.g., in the system of Reichenbach, several variants of such generalizations are introduced).

The following general statement can be formulated: among the possible functions of an *n*-valued construction there are some (at least single ones) which are generalizations of the two-valued *A* and *N*, and for any definition of 'law' ('tautology') *AxNx* remains a law; and there are also such generalizations of *A* and *N* that for any definition of 'law' *AxNx* will not be a law. Certainly the designations should be somewhat differentiated. The definition of 'law' presupposes a division of the set of truth values into two non-empty subsets, one of which corresponds to the class of asserted propositions. What has been said holds for all functions and tautologies of two-valued logic.

Of special importance is the following circumstance. Although the law of the excluded middle, in its third formulation in many-valued logic loses its absolute character (systems can be constructed where it does not count as a law), nevertheless this in no way means that its negation is accepted as a law. More than that, it does not even mean that in some cases the negation of the proposition '*x* or not-*x*' can take a value corresponding to truth, i.e. a value which is designated. In the three-valued logic of Łukasiewicz and in the logic of Heyting, for example, the negation of '*x* or not-*x*' never takes a value corresponding to truth. But even without examples, the following train of thought is sufficiently clear: admitting the negation of '*x* or not-*x*' as legitimate in some cases divests the logical constructions of their foundations – they become utterly impossible. When there seems to be an exception to this rule, then this is based on the circumstance that *A* and *N* are in fact not generalizations of the corresponding two-valued functions (a substitution occurs in the concept).

In general, if the condition is satisfied (see above) that $N1 = n$, $Nn = 1$ and $Axy = min(x,y)$, then *NAxNx* cannot be a law, since $NA1N1 = n$ (i.e., has the value which corresponds to falsity and is not asserted). As to the satisfiability of *NAxNx*, we will come back to this below (a proposition is satisfiable if it can be asserted for any truth values of the propositions out of which it is composed, of the propositional variables which it contains).

Finally, we note that if *AxNx* is a law in the sense of a tautology, then it is not always a law in the sense of an axiom. In functional constructions its being a tautology is proved on the basis of the definitions of *A* and *N*. In axiomatic constructions it is not obligatory that it be one of the axioms (as, e.g., in the constructions where only implication and negation are basic), even if it is acceptable (derivable, true, etc.) in the given axiomatic construction.

If the law of the excluded middle is formulated (is one of the formulations) not in the form *AxNx* where *A* is non-exclusive disjunction, but in the form *ExNx* where *E* is exclusive disjunction, then different kinds of many-valued generalizations can be constructed for which all our considerations about *AxNx* remain valid: in certain cases *ExNx* will be a tautology, in others not; within the limits of one and the same many-valued construction such generalizations of the two-valued functions *E* and *N* are possible, for some of which *ExNx* will be a tautology and for others not.

§ 4. The situation is similar with respect to the law of contradiction. The expression 'law of contradiction' can have at least the following meanings:

1) a proposition cannot simultaneously be true and false (true propositions are not false, false ones not true);

2) a proposition cannot simultaneously have the value i and not-i;

3) $NKxNx$, i.e., 'not-(x and not-x)'.

It is not necessary here to consider the relations between these formulations: they are analogous to the ones considered in the previous section. We add only the following remark. The first formulation remains valid in any n-valued construction (whereas the statement 'every proposition is either true or false' is not retained). With respect to the second formulation a precision can be made for the case when $x = k$ does not follow from $x = i$ (and vice versa), where $i \neq k$, i.e., where i and k are any two different truth values out of the number of possible ones: if from $x = i$ it does not follow that $x = k$ and from $x = k$ it does not follow that $x = i$, then the proposition x cannot have at the same time the values i and k.

The fate of the third formulation depends entirely on the definitions of K, N and 'tautology'. For example, in the three-valued logic of Łukasiewicz $NKxNx$ is not a tautology and in his four-valued logic it is (cf. ŁUKASIEWICZ 1951).

The law of contradiction can also be formulated in the form: proposition x and its negation Nx cannot be true simultaneously. This formulation is not necessarily a verbal expression of $NKxNx$. It can simply be accepted as a principle of reasoning without functional construction. It rather coincides with the first formulation since the reference to falsity presupposes negation. Moreover, if N is a generalization of two-valued negation, then the formulation under consideration remains valid for any n-valued construction, whereas $NKxNx$ will not always be a tautology (as in the example mentioned above).

In two-valued logic, as is well known, $AxNx = NKxNx$. This permits the laws of the excluded middle and of contradiction, taken in this form, to be considered as different notations of one and the same content. Actually even here these notations are not identical from the point of view of several interpretations, e.g., of technical ones. But in many-valued logic they can in general turn out not to be equivalent. For instance, in the system based on the Heyting matrices $NKxNx$ is a tautology, but the proposition $AxNx$ is not.

In exactly the same way, the law of the excluded middle can also be written in the form 'either proposition *x* or its negation *Nx* is true', which in a certain sense coincides with the first formulation. Let us note, incidentally, that this formulation is also not identical with the corresponding formulation of the law of contradiction: a case is possible where the law of the excluded middle is not included in the number of rules of reasoning whereas the law of contradiction is preserved. In short, formulating these laws as we have done above (i.e., according to their first formulations), their difference appears quite clearly: in this sense the law of the excluded middle is a stronger postulate than the law of contradiction.

As has already been said, *NKxNx* is not always a law in the sense of a tautology. That *NNKxNx* cannot be a law follows from the condition that *K* and *N* are generalizations of two-valued conjunction and negation. But can *NNKxNx* perhaps be satisfiable in some construction? The answer to this question, like the answer to the question of the satisfiability of *NAxNx*, needs preliminary explanations concerning negation.

§ 5. Different forms of negation in many-valued logic and their different functions have been introduced above. Here we consider negation exclusively from the point of view important for the understanding of the mutual relations between two-valued and many-valued logics. If their laws are compared, then the many-valued negations should be taken as a generalization of the two-valued one. In this generalization two points should be observed. The first consists in the following. In two-valued logic the negation of truth is falsity, and the negation of falsity, truth. In many-valued logic this should be preserved in some such form: if i is the truth value corresponding to truth, and k the one corresponding to falsity, then $Ni = k$ and $Nk = i$ should be the case. Without this condition every comparison becomes meaningless.

But that is not all. The second point consists in the following. In many-valued logic, for the determination of a tautology or for the solution of a question of the satisfiability or non-satisfiability of some propositions, the set of truth values is divided into two non-overlapping subsets of designated (asserted) values and of undesignated (unasserted, denied) values. In each of these subsets, more than one value can occur. In two-valued logic this division coincides with the division into truth and falsity. This circumstance, as has been shown in Rosser and Turquette, serves to introduce a special form of negation which transforms asserted propositions into unasserted (denied) ones and vice versa. This negation is a generalization of two-valued negation along another line. Obviously, if *M* is such a negation and *A* and *K* are generalizations of two-valued disjunction and conjunction, then *AxMx* and *MKxMx* are always asserted propositions, whereas *MAxMx* and *MMKxMx* are always denied, unsatisfiable ones.

But the crux of the problem is not this. It consists in the difficulty of observing the division of the truth values into designated and undesignated ones, when defining those forms of negation which indicate for every value the value corresponding to its negation (*M* does not do this for every value), i.e., when defining the negations as singular and not as plural functions. Let us first consider some examples before we draw general conclusions concerning this subject.

In the three-valued logic of Łukasiewicz the negation *N* is defined: $N1 = 3$, $N3 = 1$ and $N2 = 2$. By the Heyting matrices it is defined: $N1 = 3$, $N3 = 1$, $N2 = 3$. Since *1* corresponds to truth and *3* to falsity, the first condition of a generalization of two-valued negation is satisfied. The propositions *NAxNx* and *NNKxNx* are unsatisfiable in both systems. However, we thus have in Łukasiewicz $N2 = 2$ and in the Heyting matrix $N2 = 3$, i.e., it is possible that *N* does not transform undesignated values into designated ones. Thus with respect to a generalization of two-valued negation, it cannot be postulated in a general way that the negation transforms unasserted propositions into asserted ones (the question is not that the negation of this postulate be valid, but only that it be excluded from the number of conditions for the generalization of two-valued negation which have been introduced for purposes of comparing the laws of two-valued and many-valued logics).

But could one not also exclude the postulate according to which negation has to transform asserted propositions into unasserted ones, limiting oneself to the first condition of generalization? To answer this question we take the opposite way: we will explain the conditions of satisfiability and non-satisfiability of *NAxNx* and *NNKxNx* by considering these propositions themselves. Thus we will rely on the following conditions generalizing two-valued *A* and *K*: if *i* is a designated and *k* an undesignated truth value, then $Aik = i$, $Aii = i$, $Akk = k$, $Kik = k$, $Kii = i$, $Kkk = k$.

Take *NAxNx*. Let $Ni = i$; then $NAiNi = i$; and we have two possibilities: $Nk = i$ (and then $NAkNk = i$) and $Nk = k$ (and then $NAkNk = k$) Let $Ni = k$; then, if $Nk = i$, then $NAkNk = k$ and $NAiNi = k$; if $Nk = k$, then $NAkNk = k$ and $NAiNi = k$. Thus only in one case is $NAxNx$ satisfiable, namely when the negation does not transform asserted propositions into denied ones (when $Ni = i$).

Take $NNKxNx$. Let $Ni = i$; then there are two possibilities: $Nk = i$ (and then $NNKkNk = i$ and $NNKiNi = i$), and $Nk = k$ (and then $NNKkNk = k$ and $NNKiNi = i$). Suppose $Ni = k$; then there are two possibilities: $Nk = i$ (and then $NNKkNk = k$ and $NNKiNi = k$), and $Nk = k$ (and then $NNKkNk = k$ and $NNKiNi = k$). Thus $NNKxNx$ is satisfiable if $Ni = i$, i.e., if N does not transform asserted propositions into denied ones.

The first condition for a generalization of the two-valued N (postulating that the negation of the value corresponding to truth gives the value corresponding to falsity, and that the negation of the value corresponding to falsity gives the value corresponding to truth) causes no doubt. The second condition of a generalization (postulating that the negation of a designated value gives an undesignated one) needs discussion. Perhaps one is here sufficiently in agreement. In any case, one thing is clear: if this condition is accepted, then $NAxNx$ and $NNKxNx$ are not satisfiable in any construction. But if it is not accepted, then these propositions can be satisfiable.

Let us note that if this second condition is not accepted, then this does not yet mean that the possibility is excluded of a negation which transforms asserted propositions into denied ones, e.g., the first condition remains untouched.

But even the first condition alone is sufficient for showing that there is no contradiction between two-valued and many-valued logic, because, for any construction of the kind considered, it excludes the expressions $NAxNx$ and $NNKxNx$ from the number of tautologies. Acceptance of the second condition (which seems reasonable for a generalization of two-valued negation) permits drawing the stronger conclusion of the unsatisfiability of these expressions. Besides, it must be said that the negation of $NKxNx$ is precisely $NNKxNx$, and not always $KxNx$, since it is not always the case that $NNKxNx = KxNx$.

If in the logical or philosophical literature one finds $NNKxNx$ (or even $KxNx$) admitted, then this is based on an obvious misunderstanding and confusion of the concepts, and an ignoring of the postulates for a generalization of K and N.

§ 6. The comparison of two-valued and many-valued constructions is, of course, not limited to what has been said. In particular, in many-valued constructions there are functions – and consequently also tautologies formulated by means of them – which have no analogs in two-valued logic; many-valued constructions may also be devised exclusively for the survey of tautologies of a determinate kind, such that some functions generalizing certain two-valued ones may not be present in them. Further aspects of comparison can be pointed out, but they are not of as much philosophical interest as the ones considered above. Let us add some considerations connected with axiomatic constructions.

It should be noted that in functional constructions one can distinguish primitive concepts and derived ones defined by means of them; and one can distinguish primitive assertions, and assertions derivable from them according to logical rules (in particular, the definitions of the basic functions can be considered as axioms). In this case it is possible to speak of a deductive method in distinction from a descriptive one. In any case, if one speaks of the axiomatic construction of logical theory, then one means a construction different from the functional one. As has already been observed, in axiomatic constructions some sequences of symbols are defined as true, always asserted, proved, etc. (various terminologies are used) formulas (expressions, propositions, etc.); rules are indicated according to which new true formulas can be obtained from them.

It is quite obvious that the comparison of axiomatic constructions from the point of view of our subject is without meaning if they are not considered as axiomatizations of functional constructions or are not functionally interpreted. A comparison presupposes further the establishment of a correspondence between the logical symbols; more than that, it presupposes the establishment of at least a partial identity between them. We are not speaking of the fact that in one system conjunction is designated by the symbol K, in another by a point, and in a third by some other sort of symbol, but rather of the correspondence between their meanings. For example, in comparing an axiom system of classical propositional calculus with an intuitionistic one, we either consider their respective symbols of implication, conjunction, disjunction and negation as identical in meaning, or we consider the symbols of the one system as corresponding generalizations of the symbols of the other (they are, so to speak, partially identical), if we interpret the one as two-valued and the other as three- (or more) valued.

Observing the above-mentioned conditions of comparison, naturally we find from the point of view of their axiomatization no precedent for a contradiction between two-valued and many-valued logics. In particular, the absence of the law of the excluded middle from the number of axioms and derivable statements does not mean that the negation of this law occurs among the axioms and derivable statements. The situation is similar with respect to other laws of two-valued logic.

We admit a single case where propositions of the form *NAxNx*, *NNKxNx*, *KxNx*, etc. will occur as axioms of some system, namely when the class of always denied (always false) propositions has to be defined axiomatically. For instance, such a primitive system is possible:

1) *KxNx* is a derivable (provable, acceptable, etc.) formula;

2) if x and y are derivable formulas, then also *Kxy* is derivable.

But derivability should here be interpreted as ... non-derivability (deniability, unprovability, etc.), if we want to consider this system in connection with other logical theories. But such a reformulation does not change the matter in principle.

Because in many-valued functional constructions there are functions for which no analogs in two-valued logic exist, the axiomatic constructions corresponding to them can contain symbols for which no symbols with a corresponding meaning can be found in the axiomatizations of two-valued logic (cf., e.g., the axiom systems of Wajsberg-Słupecki, Rosser and Turquette, and others). This can be used as one of the criteria for distinguishing axiomatic systems.

Some properties of an axiomatic construction depend on its two-valued or many-valued interpretation. Thus, the axiom system of the classical propositional calculus can also be interpreted as the axiomatization of a many-valued functional construction. But it thus proves functionally incomplete with respect to the latter, so that tautologies formulated with the help of the symbols of functions which have no analogs in two-valued logic will not be derivable. Thus with respect to axiomatic constructions taken in their relation to functional ones, supplementary conditions can be introduced which permit one to differentiate them.

§ 7. A third group of questions concerning the mutual relations between two- and many-valued logics can be summed up in the following formulation: by what means can we obtain a many-valued logic from two-valued logic, and vice versa? As has been indicated above, there are various such means. Since, from the point of view of our subject, axiomatic constructions are taken in relation with functional ones, it will be sufficient to limit ourselves to the following characterization of these means. For example, the 'weakening' of two-valued logic by excluding some axiom can only be considered as a change to many-valued logic insofar as such a change is verified functionally: as far as the definitions of many-valued functions are constructed from the point of view where the axiom excluded will no longer be a tautology (as in the system based on the Heyting matrices *AxNx* is not a tautology). Similarly for the opposite approach.

The transition from two-valued to many-valued logic consists mainly in a generalization of the definitions of two-valued functions such that they become able to cover many-valued functions. The generalization can immediately be made for an arbitrary number of truth values, as this has been shown, e.g., in § 6 of Chapter One. But the generalization can also be achieved differently: one can show a way of obtaining many-valued definitions on the basis of two-valued ones. In § 3 of Chapter Two, for example, the method of Jaśkowski for the multiplication of matrices is set forth, which is one of the methods of generalizations of this kind. Perhaps the term 'generalization' is not fully justified here. However here, as in the case mentioned above, many-valued matrices are obtained from two-valued ones.

The transition from two-valued to many-valued logic, however, is not reducible only to such generalizations of two-valued functions. The latter is only one side, one aspect in the construction of many-valued systems:

1) thus many-valued functions will not always be obtained without exception;

2) the concept of a tautology remains undetermined;

3) the classes of equivalent expressions need not necessarily coincide with those of two-valued logic (they may or may not coincide); to make out the equivalent expressions, a supplementary investigation is necessary. In any case, at least a proof of the coincidence is needed, if it does hold.

Generally speaking, even in the case where many-valued logic appears as a generalization of two-valued logic, its construction involves a special investigation of which the generalization is only one aspect.

The situation for the transition from many-valued to two-valued logic is exactly the same. First of all, the transition to two-valued logic simply means assuming that $n = 2$ only when the n-valued construction contains no functions which are not generalizations of two-valued ones. If this condition is not satisfied, then supplementary operations are needed for the transition. For instance, for the transition from a logical system with Słupecki's function to a two-valued logic, this function itself must be disposed of: since $Tx = i$, where i is a truth value different from the values corresponding to truth and falsity, simply postulating $n = 2$ makes Tx meaningless; nevertheless a conclusion must be drawn also concerning this function. Other forms of obtaining a two-valued logic from a many-valued one have been quoted above. All these forms characteristically have the exact same need of supplementary operations besides just postulating $n = 2$. Generally, many-valued logic appears as a 'generalization' of two-valued logic (and two-valued logic a limitation of many-valued logic) only in a determinate sense of this word and only from a determinate point of view.

We might note in passing that the expression 'function having no analog in two-valued logic' needs an explanation. Take, for instance, $Tx = 2$ ($1,\ldots,n$ are truth values). If it is agreed that Tx has a truth value different from 1 and from n, then Tx is in fact impossible in two-valued logic. But taking as basic the following functions: $Bxy = i$, if $x = y = i$ ($1 \leqslant i \leqslant n$), and $Bxy = 1$ in all the other cases; $Mx = x+1$, if $x<n$, and $Mn = 1$ (cf. § 3 of Chapter Three). By their means T can be defined in the following way: $Tx = MBxMx$ (namely $x \neq Mx$; therefore $BxMx = 1$, and $MBxMx = 2$). Assuming $n = 2$, Tx thus simply proves to be an always false proposition (if n corresponds to falsity). In the case where $Tx = 3$, Tx can be defined as $MMBxMx$ and for $n = 2$ it proves to be an always true proposition, etc.

In general the following can be said: if in n-valued logic only those functions are taken as basic which are generalizations of two-valued ones (and in this sense they have analogs in two-valued logic), then all functions of this logic, being definable by means of the basic ones (expressible with the help of the symbols of basic functions), are generalizations of two-valued functions (and in this sense they all are analogs of two-valued functions). From this point of view the two-valued functions appear in their own way as classes of functions.

CHAPTER FIVE

THE PROBLEM OF THE INTERPRETATION OF MANY-VALUED SYSTEMS

§ 1. The question of the interpretation of many-valued logical systems is connected in an organic way with the question of the epistemological nature of these systems (in the sense: do they reflect properties of reality ordo they not?) and with the understanding of the truth values themselves. Lo gical systems, as they occur in the context of the development of knowledge as a whole, are constructed not simply to satisfy the curiosity of some group of people, but rather for the purpose of using them to solve concrete scientific problems, and the question 'do they reflect something or not?' is thus unavoidable. And if we want to apply logic (or if we want to examine the question of the possibility of its application), then it is absolutely necessary to explain the meaning of the terms 'true', 'false', etc. (cf. HEYTING 1956a), or in general the symbols designating truth values. If it is still to some extent possible in two-valued logic to refer to the intuitive clarity, usage, etc. of these terms (although even here the situation is not as simple as it seems at first glance), then such references are impossible in n-valued logic, what, for example, is the meaning of the truth value to which the number *20* is assigned? This question cannot be answered without a special convention.

Although many, if not most, of the authors of works on many-valued logic try to avoid an interpretation of their constructions, nevertheless enough factual material of different kinds has accumulated along these lines, that on its basis the problem of the interpretation of many-valued logical systems can be examined in a general way.

The interpretations available are quite varied not only with respect to their object domain – modal propositions, normative propositions (cf., e.g., KALINOWSKI), relay switching circuits (Šestakov, Jablonskij), mathematical reasoning (Bočvar, Heyting matrices, constructivistic logic), the language of quantum mechanics (Reichenbach), the phenomena of the microcosmos (Destouches-Février), the theory of probability (Łukasiewicz, Reichenbach), the problem of entailment (Lewis and Langford, Ackermann), etc. – but also with respect to some of their general features. Obviously a onesided approach to this question is hardly profitable.

We consider here two basic types of interpretations, whose interest is, from a philosophical point of view, entirely different.

By 'the interpretation of a many-valued logical system' we mean, first of all, the establishment of a correspondence between the elements of the logical system and a domain of concrete objects; between the propositional variables, their truth values and the propositional functions on the one hand, and the objects, their properties and mutual relations on the other hand. In this case the logical system plays the rôle of a model (in the sense of ZINOV'EV and REVZIN) for a given object domain and is an imitation of the latter. As far as the object domain remains undifferentiated here (in the sense that it can be a language system, a technical system, the behaviour of micro-particles, etc.), the meaning of the symbols of the truth values is also undifferentiated (here only a generalization, about which we will speak below, is possible). Therefore the truth values are here normally defined as a set of symbols (normally: of numbers); it is an idle question as to which meaning each of these symbols has by itself, independently of possible interpretations and of the conveniences of the calculi.

The following example shows quite clearly that the terms 'true', 'false', 'undetermined', etc., are here not defined, but are only replaced by terms from a concrete domain. We make a contact a correspond to the proposition x, and one of the contact's possible positions i (it is indifferent which one) to truth. Under this interpretation, the proposition 'x is true' has as a translation in the language of the given object domain the expression 'the contact a is in position i'. Thus it is immaterial which truth value we let correspond to the position i: we could as well have chosen falsity.

By 'the interpretation of many-valued logical systems', we mean secondly, a use made of them which is based on a definition of the truth values (for the present this concerns three-valued logics) where it is not the concrete nature of some object domain which is taken into account, but where the terms (the symbols in general) are defined as concepts characterizing the process of knowledge. (For the distinction of these two types of interpretation cf. also ZINOV'EV 1959e). It would be inexact to see the distinction of these interpretations as a distinction between different object domains: physical objects on the one hand, propositions on the other. We will give an example below of an interpretation of a many-valued system in terms describing propositions, but this will be an interpretation of the first type, whereas the approach of Łukasiewicz and Tarski to the same domain can be related to the interpretations of the second type. The difference here is something else, namely the one described above (this will become clearer by further examples). Of course, this distinction is not always sharp and absolute (thus, for example, in comparing the works of Reichenbach and Destouches-Février, this distinction can only be found if several abstractions are made, i.e., if the comparison is one-sided), but nevertheless it actually exists. For instance, take the proposition 'contact *a* is closed'. It is defined as true if in fact the contact *a* is closed. Instead of the term 'true' any other term may be used, but the situation remains the same: it concerns the relation of the proposition to reality. Similarly the proposition 'proposition *x* is true' is true, if in fact *x* is true. Evidently we have here no translation like in the case of the first type.

Within the limits of both types of interpretation mentioned, a gradation can be introduced. But it is not possible to do this here, so we shall limit ourselves to giving some examples which will clarify what has been said above.

§ 2. Let us first turn to the logic of modal propositions. Not that we intend to take up here the investigation of these propositions, although it is closely related to many-valued logic' (cf., e.g., VON WRIGHT, ACKERMANN). As an illustration of the general situation, we will give an exposition of a rather simple three-valued formal system, which can be interpreted as a theory of modal propositions (GRENIEWSKI 1955). This system is based on the fundamental concepts of the logic of Łukasiewicz. Of course, by means of it only a one-sided and partial description of modal propositions can be given. Thus we do not rule out the possibility of investigating them in a two-valued logic or in any other many-valued system (e.g. in a six-valued one; cf. ACKERMANN). This means – a circumstance which merits some attention – that for the solution of at least some scientific problems, many-valued logic or some determinate variant of it are not absolutely necessary but represent only one among several possible means.

The criterion for the efficiency of these means is in the end the practical value of the theoretical results obtained.

Thus, we take the following expressions as primitive:
1) $0, \frac{1}{2}, 1$ – truth values,
2) $x, y, z, \ldots$ – atomic propositions.

The operator N is defined as in Łukasiewicz, and the operators N^1 and N^2 we define by the matrices:

x	N^1x	N^2x
0	0	1
$\frac{1}{2}$	1	0
1	1	0

By their means N^3 and N^4 are defined: $N^3x = N^2Nx$, $N^4x = N^1Nx$. We define the operator C as in Łukasiewicz, and by means of it A and K (as it has been done above), and also R: $Rxy = KCxyCyx$. An asserted proposition being defined as a proposition taking the value 1 for any values of the atomic propositions out of which it is composed, we can distinguish a whole set of propositions, formulated by means of the symbols of atomic propositions and the operators introduced, which contains the propositions asserted in the given system; e.g. Cxx, AN^2xN^1x, AN^3NxN^3x, etc. (cf. GRENIEWSKI 1955).

Further, if we interpret Nx as 'not-x', N^1x as 'it is possible that x', N^2x as 'it is impossible that x', N^3x as 'it is necessary that x', N^4x as 'it is not necessary that x', then the asserted statements of the given system and the definitions will describe the properties of the modal functors and the mutual relations between the propositions containing them. In particular, the definition of N^3 will then read: 'that it is necessary that x, means that it is impossible that not-x'; and the assertion AN^2xN^1x will read: 'it is impossible that x or it is possible that x'; etc.

As we see, the formal construction (in the example we necessarily have simplified it) of many-valued systems, and their use by giving them an interpretation in terms of a determinate object domain, in no way differs, from a philosophical point of view, from the same procedures in two-valued logic. Owing to the fact that one is dealing with an arbitrary number of truth values, new problems originate. But they all arise on another level, namely on the level of the relation between many-valued and two-valued logics, on the level of the informal understanding of the supplementary truth values and on the level of the nature of logic in general.

In the example considered, the many-valued construction plays the rôle of a model for the modal propositions (for the concepts of model and interpretation cf. ZINOV'EV and REVZIN, where the necessity of differentiating these concepts is treated). Thus the use of this model (of this interpretation of the formal construction) is not made possible by giving definitions for the truth values, but by establishing a correspondence between the symbols N, N^1, N^2, etc., on the one hand, and the negation 'not' and the modal functors 'possible', 'impossible', etc., on the other hand. In other cases the interpretation is obtained by the establishment of a correspondence between the symbols of the truth values and possible states of the objects (in the theory of electrical networks), but also this does not mean a definition of the truth values, since in each case another signification is given to them.

Thus, if we go over to systems of discrete activity, the truth values will be made to correspond to the states of the elements of these systems, and the propositions will correspond to the elements themselves. In JABLONSKIJ, for example, such an illustration is mentioned: suppose some system is given which has k inputs and l outputs whereby each can be in one of a finite number of states; the workings of the system can be described by the functions $F^1(x^1,\ldots,x^k),\ldots,F^l(x^1,\ldots,x^k)$, which prove to be functions of a many-valued propositional calculus.

Interpretations of the second type are given in the works of Łukasiewicz-Tarski, Bočvar, Reichenbach, Šestakov, Kleene (the works of Šestakov are clearly examples of interpretations of the first kind, but in ŠESTAKOV 1960, there is also an exposition of the considerations of Bočvar and Kleene) as already mentioned in the preceding chapters. In these cases ordinary truth and falsity are considered, and the third value is defined as 'possibility' or 'neutrality', as 'indeterminateness' (the proposition can neither be verified nor falsified), 'meaninglessness' (a proposition having no meaning can as such belong neither to the true nor to the false propositions), 'insignificance' (it is not important if it is true or false), 'algorithmic undecidability', etc.

As we shall see, the interpretation of the truth values here is not connected with a concrete object domain, but definitions are given.

Depending on works of Łukasiewicz, Tarski gave the following definition of possibility (cf. ŁUKASIEWICZ 1930): $Mx = CNxx$ where C and N are the three-valued operators of Łukasiewicz. Verification shows that $M0 = 0$, $M\frac{1}{2} = 1$, $M1 = 1$. On the basis of this definition the consistency of the following assertions can be shown: $CNMxNx$, $CNxNMx$, $(\Sigma x)KMxMNx$. E.g.: $CNM0N0 = CN0N0 = C11 = 1$, $CNM\frac{1}{2}N\frac{1}{2} = CN1N\frac{1}{2} = C0\,\frac{1}{2} = 1$, $CNM1N1 = CN1N1 = C00 = 1$; $(\Sigma x)KMxMNx = (KM0MN0)A(KM\frac{1}{2}MN\frac{1}{2})\ A(KM1MN1) = (K01)$ $A(K11)A(K10) = 0A1A0 = 1$. As we see, no correspondence of the type of the first interpretation is established here. Mx is here defined within the limits of three-valued logic and the truth values assigned to Mx depend on the values of x as follows:

x	Mx
0	0
$\frac{1}{2}$	1
1	1

Let us give a further example (taken from JABLONSKIJ) which shows that the types of interpretation considered have points of contact. Suppose some closed interval is divided into n equal parts. We consider the position on it of a particle having commensurable dimensions. The proposition xi ($1 \leqslant i \leqslant n$) states that the particle is located within the interval with the number i. We take the assertions: 1) xi is true, if the particle is entirely inside the interval with the number i; 2) xi is false, if the particle is completely outside the interval with the number i; 3) xi is undetermined in all the other cases (i.e., when the particle occupies points in two neighboring intervals or occupies a limiting point of an interval).

This example can be considered as an example of both types of interpretation: of the first type, if the assertions made are simply considered as establishing a correspondence between the symbols 'true', 'false' and 'undetermined' on the one hand, and possible positions of the particle on the other hand; and of the second type, if the terms 'true' and 'false' are used to characterize in a general way the relation of propositions to reality, for instance, in the following way: truth and falsity relate the proposition to two different possible states of the object, and indeterminacy to a transitional state. A similar example, to which we already have referred, is treated in ROSSER and TURQUETTE.

Above we have noted that many authors of works in the field of many-valued logic refrain entirely from referring to the interpretation of many-valued systems. And this is no accident. The reason certainly is not that it would be difficult to find examples – this for once is not hard. The reason lies above all in the circumstance that, first of all, an interpretation is understood not simply as an interpretation of the many-valued propositions taken separately (such an interpretation would amount to an intuitive example for common sense), but as an interpretation of the calculus as a whole; and, secondly, that it is understood as a means for the solution of concrete scientific problems. But a proof of the efficiency of many-valued constructions and of their superiority over two-valued constructions is something which needs time. From this point of view, many-valued logic represents no fatal, blind necessity, but only one of the levers consciously applied in the scientific process. And it must be admitted that its elaboration as a formal apparatus is the more important task, which, of course, does not exclude the search for applications.

The facts about the interpretation of many-valued systems give a quite convincing answer to the question whether or not these systems reflect properties, relations and connections of things: an affirmative answer. If one only admits that many-valued logic describes the language of quantum mechanics, then the question remains: but why is such a logic convenient and not another one? Obviously, because in science just those conditions of knowledge have occurred which made the use of three-valued logic possible, and these conditions are an objective fact.

§ 3. The interpretations of the second type are no doubt of greater interest for philosophy than those of the first type: it is here that the definition of the truth values is discussed.

Take the following example (cf. DESTOUCHES-FÉVRIER). Suppose that on the scale of some measuring instrument the divisions *1*, *2* and *3* are marked, and that the pointer can point to exactly one of these divisions (for simplicity we exclude all intermediate positions). We take, further, the following propositions:

1) the pointer points to division *1*;
2) the pointer points to division *4*;
3) the pointer points to one of the divisions *1*, *2* and *3*.

The first is true, if the pointer points to *1*, and it is false, if the pointer points to *2* or *3*. Either one can be the case. The second proposition, however, will never be true, because there is no such division at all on the scale of the instrument. The third proposition will always be true, because the pointer points necessarily to one of *1*, *2* and *3*. We see, in the given case the truth value of the propositions is defined by their confrontation with the reality which they reflect; four cases are possible, which can be called: 'absolutely true' (third proposition), 'absolutely false' (second proposition), 'possibly true' and 'possibly false' (first proposition). The terminology chosen is not essential. What is important is the general principle of the definition, which already leads us into the domain of logical semantics.

A second example: suppose $P(a)$ symbolizes a proposition affirming that some object (called a) has some property (called P), say 'contact a is closed', 'number a is an integer', 'number a is not even', etc. We will confront such propositions with some given situation. First of all the question arises: does object a occur in the given situation? If it does not, how is the proposition to be evaluated? Obviously, it cannot be evaluated as true, nor as false. It is meaningless for the given situation, undetermined, etc. In any case, its value is different from truth and falsity, if the latter are understood in the following way: the proposition is true, if in the given situation a actually has the property P; the proposition is false, if in the given situation a does occur but it lacks the property P. Of course, falsity could also be defined as follows: $P(a)$ is false, if $NP(a)$ is true meaning that a does not have the property P; in this sense the proposition 'number a is an integer' is false, if it is true that number a is not an integer.

A third example: take a proposition of the form 'if x, then y', expressing a connection between different objects. This proposition can be confronted with one of the possible situations described by the propositions 'x and y', 'not-x and y', 'x and not-y' and 'not-x and not-y'. Obviously, here also several (more than two) evaluations called truth values are possible.

Independently of the differences of the given examples (their number could without difficulty be enlarged), the following holds for them all:

1) the propositions are confronted with the situation described (hypothetically) by them, or in general with some situation from some domain of reality, if the proposition is constructed out of other propositions and defined as a function of the latter, it also is finally confronted with some situation, namely through confrontation with the situations of

the propositions composing it taken in a determinate order (or in random order, which is a particular case);

2) the confrontation takes place as an ordered procedure, depending on the structure of the proposition;

3) with regard to the propositions, in this procedure at least two steps are possible (simple propositions, propositions with negations included, have at least a subject and a predicate; complex propositions consist of or are obtained from at least two propositions); this means that at least four different evaluations of the proposition in its relation to a given situation are possible (in general 2^n evaluations where n is the number of steps);

4) already within the limits of ordinary language one can have recourse to a n-valued evaluation of propositions;

5) all n values of a proposition can be defined by the truth of other propositions, and truth can be defined as correspondence between the proposition and the situation with which it is confronted.

From what has been said it does not follow at all, that it is necessary to proceed thus with respect to ordinary language and the language of some concrete science. This is all only a possibility, for the realization of which we need important reasons (cf. Bočvar, Reichenbach, *et al.*). In the overwhelming majority of cases it is sufficient to know if the proposition corresponds to reality or not. And the character of the correspondence or non-correspondence (their different gradations) is established by verification, so that no special terminology is needed which might complicate the languages. Thus, if it appears that object a does not exist in the given situation or does not exist at all, then the indication of this fact in the evaluation of a proposition $P(a)$, which is different from the proposition 'a does not exist', will either be equivalent with the evaluation of $P(a)$ as false in case $P(a)$ is the proposition 'a exists', or else it will give $P(a)$ some kind of third value if nothing is said concerning existence.

The most important thing discovered in this way is perhaps the fact that all n-1 values can be defined by means of a single one, namely by means of truth. In this manner, many-valuedness appears here as a consequence of the one-valuedness of knowledge (potentially or in tendency). But many-valued logic as specifically many-valued appears precisely when n different truth values are presupposed, and when, on basis of this presupposition, a logical system is constructed which considers the connections between the many-valued propositions.

§ 4. Understanding the truth values of propositions on the condition that any finite or denumerably infinite set of such values is admitted seems to be one of the central philosophical problems of contemporary logic. We recall that what is here in question is not the interpretation of the expression 'proposition x has the truth value i' in terms of some concrete branch of science, but the general definition of the conditions under which the proposition x is evaluated as having the truth value i.

The actual points of view concerning this subject, which we have occasionally mentioned, can be divided into two groups (cf. for this ZINOV'EV 1959 e). To the first group belong those interpretations of the supplementary truth values besides 'truth' and 'falsity', according to which these values are made dependent on the realization, the realizability or the importance of the verification of the propositions. Thus only three-valued logic is covered where the third value is defined in the following way: it is unknown (it cannot be established, is indifferent, etc.) whether the proposition is true or false. Moreover, in such an approach the character of the proposition remains unclear: it may appear that it either has one of the two truth values, or that it is true under some conditions and false under others, or that under one and the same condition it is neither true nor false because of the very definitions of 'truth' and 'falsity', etc.

To the second group belong those interpretations according to which the truth values are considered as independent of the realization, realizability or importance of the verification of the propositions. Different variants of this group have also been enumerated in ZINOV'EV 1959e. In the majority of them the truth values are reduced to probability or modality, so that the problem of the supplementary truth values loses its special character. In several works of this group (of Bočvar, Reichenbach, *et al.*) the third truth value is not reduced to truth and falsity, but also to modality and probability: if a proposition has the third truth value, then it is neither true nor false. However, in these works the authors also understand the third value as meaninglessness, as impossibility of confirmation or refutation, etc., which is still far from satisfying the postulate of the generality of the definitions.

It is characteristic that in the overwhelming majority of cases the propositions are taken as something elementary, not analyzable into parts. And even in the cases where the structure of the propositions is mentioned, its impact on the explanation of the truth values and on the introduction of their definitions is not realized. Thus in the work of Reichenbach, among the undetermined propositions are included those about unobservable objects and therefore not subject to verification (if an object cannot be observed, then it is impossible to decide whether it has some property or does not have it, and in the work of Rosser and Turquette, in considering the foundations of quantification in many-valued logic, the possibility is admitted that the proposition *P(a, b,...)* can take *n* truth values, depending on the values of the variables. In the first case indirectly, in the second case openly, the structure of the propositions is touched upon, however it is not connected with the possibility of truth values besides the classical ones (in the second case, the truth values are interpreted by means of a classification of the values of the variables).

We do not claim that taking into account the structure of the propositions is the only way to an understanding of the truth values. On the contrary, the greater the variety of investigations in this respect, the wider will be the possibilities of making use of the ideas and the apparatus of many-valued logic. However, it would be expedient to use this way also as one of the possible ones. A realization of this task is connected with the solution of several difficulties concerning the elaboration of a general theory of symbols, of a theory of the correspondence relations, etc., because the propositions are thus taken as sets of symbols of different kinds, ordered in a definite way, and the process of establishing the character of their correspondence to reality (their truth values) is an ordered procedure confronting the elements of the propositions with objects, properties, relations and connections of the given situation. We are not concerned here with the empirical verification of propositions, but with the question of principle of a schema of definitions which would allow us to draw conclusions of the type: that x has the truth value i, means, according to the definition of i, that y is true. Strictly speaking, this is only a further elaboration of the two-valued schema for falsity: x is false means, according to the definition, that Nx is true.

It would follow perhaps, from the interpretations of many-valued logical systems, that one should distinguish that use of them which itself is needed in some interpretation. The example given of the definition of Mx as $CNxx$ is characteristic in this respect. This definition allows one to establish the consistency of some intuitive prescripts for Mx and to define Mx as a three-valued function of x. However, the three-valuedness itself thus remains without interpretation. What is said is actually related to the use of many-valued logic for the solution of problems of this logic itself: a question which needs special consideration.

§ 5. An analysis of the facts concerning the interpretation of many-valued systems shows that the latter appear as really many-valued in the strict sense of the word only when they are used as an apparatus for the investigation of connections between objects. Naturally, the representation of many-valued functional constructions (and possibly also of the axiomatic ones corresponding to them) as parts of the theory of the connections between objects (cf. ZINOV'EV 1959c, 1959d), is of some interest.

For this it is necessary to give the following interpretation to the propositions and truth values:

1) the symbols of propositions are considered as symbols of objects;

2) the symbols of truth values are considered as symbols of possible properties (or states) of the objects;

3) the expressions of the form '$x = i$' must be read like expressions of the form 'object x has the property i';

4) all other expressions are interpreted according to the definitions; e.g., Nx depending on $Nx = \alpha$, where α is the defining part, will designate a combination of properties other than x. Of course, for such an interpretation the introduction of several preliminary concepts ('object', 'property', etc.) is necessary. Such an interpretation has many merits leading toward an elimination of the ambiguity of several concepts, because the calculus is constructed in a very general terminology which admits as one of its applications the particular case of the connections between propositions.

In such an interpretation it is convenient, furthermore, to consider complex propositions composed of atomic ones by means of the operators $A, K, C, \ldots$, as objects different from the objects designated by the symbols of atomic propositions, and to introduce the concept of an ordering of the objects. For instance, instead of the definition $Kxy = min(x,y)$ the definition $z = min(x,y)$ can be introduced which is concerned on the one hand with a connection between x and y and on the other with z. The ordering of x, y and z appears in the fact that (in the example) the definition can be read in two directions:

1) if x and y have such-and-such a property (are in such-and-such states; appear as combinations), then z has such-and-such a property;

2) if z has such-and-such a property, then x and y have such-and-such properties, i.e. (in the example) $min(x,y) = z$. The unimportance of the order of x and y is a particular case of ordering, which in every case is agreed upon. In other words, the ordering of x, y and z is expressed by the order of reading the tables

a)

x	y	z

b)

z	x	y

The example considered is an example of a complex connection (a connection between three objects). A simpler case of a connection is a connection between two objects, which is described by a function of one argument. Thereby Fx is considered as an object different from x. In a table this has the form:

x	y
1	α^1
2	α^2
⋮	⋮
n	α^n

where each of $\alpha^1, \alpha^2, \ldots, \alpha^n$ is equal to one of the numbers $1, 2, \ldots, n$; moreover, it is possible that $\alpha^i = \alpha^k$ $(1 \leqslant i \leqslant n,\ 1 \leqslant k \leqslant n)$.

If Fx is understood in this way, then it is convenient, of course, to introduce supplementary symbols of negation. For example, the negation N^* can be understood as a renumbering of $\alpha^1, \alpha^2, \ldots, \alpha^n$ such that $N^*x = x+1$, and if $x = n$, then $N^*x = 1$. It is necessary to introduce also the negations of connections on account of the ordering of the objects. For example, the following definition is possible: 'it is not true that if x then y = if x then not-y.'

The definitions of the functions of the propositional calculus have the task of showing what truth values some propositions take depending on the truth values of others. In the given interpretation, they will define which of the possible strictly verifiable properties (states) some objects will take depending on the given properties of other objects. The fact that logical calculi reflect definite properties of things is quite evident in this interpretation.

Finally, in such an interpretation it is convenient to distinguish the precision of the logical connectives 'and', 'or', 'if...then...', etc., as realized in the propositional calculus (in functional constructions) from that which can be realized without interpretation in a functional construction (so to speak, by way of meaning): if the functions of propositional calculus will be interpreted as the description of connections between objects, then the language with whose help this description has to be realized, should be made precise.

In ZINOV'EV 1959c and 1959d the following method is proposed regarding this:

First stage: we define in some axiom system the basic logical connectives 'and' ('each of the ones which are spoken about'), 'or' (exclusive 'or'), 'not' and 'signifies' ('consequently') (cf. KOLMOGOROV 1925). Thus the axiom system should give the rules for operating with these connectives and the connectives should preliminarily be explained by examples, by way of equations, etc.; in general, operating with them should become a habit. For example, among this kind of axiom can occur axioms of the form: C^*K^*xyx, $C^*K^*A^*xyzA^*K^*xzK^*yz$, $C^*N^*A^*xyA^*K^*xyK^*N^*xN^*y$ etc., where K^*, A^*, N^* and C^* are respectively 'and', 'or', 'not' and 'signifies'. The meaning of this kind of axiom is easily explained: e.g., if we say x and y, then by the same token we say x. The difficulties with the symbol 'signifies' are (in the given case, of course) very small: it can be understood thus, that there are some rules for writing immediately after an inscription some other inscription defined by these rules. The axioms, properly, also give such rules. We note, finally, that the axiom system as a whole should define these symbols (in the example all axioms are elements of the definition of all symbols used).

Second stage: the symbolic language constructed in the first stage (it differs from ordinary language only by its brevity, lack of ambiguity, operability) is used for the construction of a general theory of connections, in relation to which the theory of the functions of propositional calculus is only a fragment. In this case, the operators N, K, A, C and others already appear not as explications of logical symbols of ordinary language, but in a new rôle – the rôle of symbols for different kinds of connections. A novelty here, in particular, is the fact that in the matrices, as well as in the equations of truth values corresponding to them, used for the definition of these operators, a consideration of order appears, once we distinguish the defining and the defined parts (the arguments and the functions in general). For instance, writing down $Axy = min(x,y)$, we show that Axy takes such-and-such a value depending on the values x and y, and not vice versa.

From this point of view the operators considered can be defined as follows:

1) if $z = min(x,y)$, then $A(x,y)z$;
2) if $z = max(x,y)$, then $K(x,y)z$;
3) if $z = max(1, y—x+1)$, then $C(x,y)z$;
4) if $z = n—x+1$, then Nxz;
5) if $z = x$, then Rxz;

etc. (if x, y, z,... are propositions, then 1, 2,..., n are their truth values; if x, y, z,... are symbols for objects in general, then 1, 2,..., n are their possible states).

If it now becomes possible (to a certain degree) to replace z by Axy, Kxy, Cxy, Nx, etc., in the sense of A^*xy, K^*xy, C^*xy, N^*x, etc., then this does not make any essential change. The interpretation considered cannot, of course, pretend to uniqueness and to a fundamentally privileged position, but to us it seems to explain some new properties of contemporary logic as compared with the traditional one, and in particular, some aspects of many-valued logic. Thus the words of Hilbert concerning the predicate calculus (cf. HILBERT and ACKERMANN) can also be referred (and perhaps, primarily) to the propositional calculus which lies at its basis: Aristotelian formalism would not suffice, where the symbolic representation of the relations between several objects is concerned (and connections are a particular case of such relations).

§ 6. In connection with the interpretation of many-valued systems, criticism of the principles of two-valued logic from the point of view of dialectics is of obvious interest. We did not count dialectics among the sources of many-valued logic because it did transcend (and does transcend) the framework of formal logic in general, and therefore did not have any visible influence on the growth of many-valued logic. The term 'formal logic' is used here solely because dialectics in general, or one or another of its parts, is often spoken of as 'dialectical logic'. On the other hand, what has been said before is sufficient to establish the following conclusion: that the growth of many-valued logic did not in any way change logic (i.e. formal logic) into dialectics or a part of dialectics (i.e., into dialectical logic).

The criticism of the principles of two-valued logic in dialectics consisted essentially

1) of explaining the insufficiency of formal logic in general with respect to the investigation of complex systems of relations which arise and develop historically (e.g. of social systems), and

2) of explaining the unsuitability of formal logic as a methodology of such investigations.

An affirmative answer in this case certainly did not signify a negation of the principles of formal logic. It meant only one thing: namely, the observance of these principles presupposes as a necessary condition a definite methodology of the process of investigation. In particular (and this interpretation appears clearest here), the numerous antinomies arising in political economy were removed not simply by making the actual knowledge of bourgeois economics conform to the law of contradiction and the law of the excluded middle, but required a radical rebuilding of the entire process of investigation of the subject, which had the additional consequence of removing situations which were logically contradictory (like the antinomy of values, the antinomy of the rise of profits, etc.).

In investigating those systems of relations which develop historically and vary in many different ways, one must have recourse to various methodical procedures necessary to construct theories of these systems: some relations have to be excluded and others have to be studied; the knowledge obtained in this way has to be unified according to definite logical rules in order to give a more or less complete and approximate theoretical picture of the reciprocal actions of these relations; the contrary influences on the objects have to be considered; and the change of the system and of its structural components in time, etc., has to be checked (cf. in this connection ZINOV'EV 1958, 1959a, 1959b). Naturally, one thereby comes across propositions which, taken out of their context, contradict one another. But such propositions cannot be measured by logic if one ignores the whole systematic knowledge by means of which they have been obtained and in reference to which they make sense. An investigation shows that in this case the conditions for the application of the principles of logic are lacking: either the condition of identity in time is not fulfilled, or there is no identity with respect to the sense of the concepts, or the sense of the propositions depends on different situations (e.g., a proposition x may in fact be part of a proposition 'if x, then y', and Nx may be part of a proposition 'if Ny, then Nx', etc.). In this way the conflicts with formal logic prove imaginary: its principles are not rejected; but the question does arise concerning the correct course of the complex process of investigation.

Evaluating the propositions with respect to their truth values by using only the concepts 'true' and 'false' proves to be insufficient: one has to speak here of a larger or smaller degree of abstraction and concreteness, of approximation, completeness, etc. of the propositions – and all this means evaluation of the propositions in relation to reality. But the above-mentioned many-valuedness of the propositions needs no many-valued logical theory for its description and comprehension: all the indicated cases (and others) concerning the evaluation of the truth value of propositions can be fully comprehended by the concepts 'true' and 'false' together with supplementary significant assertions, which in general may take the form: 'proposition x is true (or false) under the condition that y'.

Attention should be paid to the fact that the possibility of interpreting (and such a possibility is not excluded) the truth values of many-valued logic as degrees of abstraction, concreteness, approximation, relativity, etc., quickly becomes an abstract possibility for the following reason. When one speaks of the many-valued character of propositions, it is presupposed that one and the same proposition x can have a value out of the number of possible values *1, 2,…, n*. But here one is considering different propositions – or more exactly – relations between different propositions (differing according to their content). It is especially important to take this circumstance into account in order to avoid fruitless attempts to represent the course of investigation under consideration as a logical calculus, and in order to avoid the confusion resulting from operating with the concepts in an uncontrolled fashion. The point is that when the propositions are evaluated as abstract, concrete, more abstract, more concrete, more complete (in the sense of a more complete description of the object), etc., then one is always considering the relations of different propositions, differing in their content and in the way they have been obtained. E.g., the Clapeyron-Clausius equation describes the state of a gas more completely, more exactly than Gay-Lussac's Law (generally speaking: it is more concrete), and Van der Waals' equation is more concrete than the Clapeyron-Clausius equation. Obviously, one is talking here about the stages in the course of the investigation and the reasoning, and not of the possible truth values of propositions taken in isolation.

The situation is different when the passing stages of the objects themselves, the very act of their change, is underconsideration. In this case, of course, to answer the questions following the 'yes – no' principle appears to be somewhat rough and gives no exact description of reality. We note, by the way, that such facts are mentioned in several works on many-valued logic. Thus in ROSSER and TURQUETTE, the simple example of a man entering a room is analyzed in detail, and it is shown what logical difficulties result when such a transitional stage is subjected to the principles of two-valued logic.

Still greater difficulties arise from the well-known paradox of Zeno, about which discussion is still going on. The author's point of view in this question is expounded in detail in ZINOV'EV 1959b. Briefly, its essentials can be summed up as follows:

1) the general form of Zeno's paradox can be expressed by the proposition: 'A changing object at the same time both has some property and does not have it';

2) this proposition is derived by way of considerations based on logical principles, and is not the result of an experiment or an observation (why this is so becomes clear in the following point);

3) it does not contradict the laws of logic, because there is a different meaning to the concept 'at the same time' which appears in Zeno's proposition and in the laws of logic (if they are, say, formulated in the form 'an object cannot at the same time have and not have some property', etc.); the first case is concerned with a time interval not equal to zero, and the second case with an instant of time, i.e., with a limit of two time intervals; as its dimension is equal to zero, the truth value of the proposition cannot be established from the point of view of practical control.

It is, therefore, possible to consider the situation here from the point of view of three values: true, false and undetermined. One can also limit oneself to stating the fact that one is concerned with a purely logical consideration referring to a zero interval. But another side of the matter is more important for us: the 'paradox' of change and the reluctance to gain an understanding of the antinomies which arise in the investigation of complex and changing reality, led to the tendency, dating from Hegel, to criticize classical logic by opposing to it the assumption of simultaneous truth and falsity for a set of privileged propositions. Insofar as the classical relation of truth and falsity is thereby preserved, this assumption is equivalent to the assumption that the proposition is neither true nor false. In general, propositions of the type $KxNx$ are in several cases considered acceptable. Various reactionaries who maintain such a criticism can still be found.

Attempts have been made in the following way to reconcile the aforementioned assumption with the law of contradiction and the law of the excluded middle: although in several cases $KxNx$ is legitimate, nevertheless even in these cases the laws of logic are retained, e.g. in the form $NKKxNxNKxNx$. But such a solution of the problem is clearly illusory.

In formal logic the assumption under consideration is not acceptable: insofar as falsity is the negation of truth, and truth the negation of falsity, this assumption signifies that in several cases true propositions prove to be not-true (false), and false propositions prove to be not-false (true). Operating with only the concepts of truth and falsity in the classical sense, in the framework of logic one can permit only those situations in which it is impossible to verify or prove whether the proposition is true or false, or in which the truth value is not essential. But in such a case the third value excludes the two others: if $x = 3$, then $x \neq 1$ and $x \neq 2$, whereas in the opposite case, if $x = 3$, then $x = 1$ and $x = 2$ $(x = 1 = 2)$.

The attempt to interpret the third value as the conjunction of truth and falsity (of assertion and negation) leads to a curious result when one tries to construct a corresponding calculus.

The assumption that in several cases $KxNx$ is acceptable can be written in the form $(\Sigma x)KxNx$, which reads: there is an x such that the conjunction of x and not-x is true. In terms of truth values this is expressed by: $(\Sigma x)KxNx = (K1N1)A \ldots A(KnNn)$, and in the case of three values $(\Sigma x)KxNx = (K1N1)A(K2N2)A(K3N3)$, where A is the symbol for disjunction.

Interpreting K as a precise formulation of the ordinary 'and' and assuming the value 1 for 'truth' we ask the following question: when will $(\Sigma x)KxNx$ be true? As far as $KxNx = max(x, Nx)$, $(\Sigma x)KxNx = 1$ only in the case where $K1N1 = 1$, because $K2N2 \geqslant 2$ and $K3N3 \geqslant 3$. But $K1N1 = 1$ is only possible if $N1 = 1$. If $N1 = 2$ or $N1 = 3$, then correspondingly $K1N1 = 2$ or $K1N1 = 3$. Thus, only one case which satisfies the assumption under consideration is possible, the case where the negation of truth is truth, and this disagrees with the very notion of negation.

Of course, a construction is possible in which the third value is the conjunction of the remaining two. For instance, if from $x = 1$ follows $x = 2$, or from $x = 2$ follows $x = 1$, i.e., $x1{,}2 = ACx1x2Cx2x1$ we get

$$\frac{\begin{matrix}Cx1x2\\ x1\end{matrix}}{x2} \text{ and } \frac{\begin{matrix}Cx1x1\\ x1\end{matrix}}{x1}, \text{ or } \frac{\begin{matrix}Cx2x1\\ x2\end{matrix}}{x1} \text{ and } \frac{\begin{matrix}Cx2x2\\ x2\end{matrix}}{x2},$$

i.e., $x = 1 = 2$. But interpreting in this case *1* as truth and *2* as falsity, we get the result that falsity follows from truth, and truth follows from falsity – i.e., we get the ordinary paradoxical situation.

There is no necessity to analyze either further possibilities for solution or other aspects of the problem. From what has been said a single conclusion can be drawn, a conclusion which the specialists in dialectics themselves recognize in the majority of cases as quite exact: dialectics by its very essence cannot be subjected to formalization by way of a formal calculus; it has other tasks than logic, and solves them by other methods. All attempts to construct it as a calculus can only lead to paradoxical situations. This, of course, does not exclude the possibility that in the solution of concrete problems dialectics will make use of the ideas and principles of logic, particularly of logical models.

From what has been said, one has to conclude that it would be wrong to think that the criticism of formal logic from the point of view of dialectics belonged to the line of thinking of many-valued logic. Dialectics did actually point to the fact that two-valued logic has limitations, once it excludes situations of the kind $(\Sigma x)KxNx$. But this limitation is one which makes science viable.

Many-valued logic destroys the philosophical illusion of the absolute and *a priori* character of two-valued logic. And in this regard, its influence on philosophy concurs with the influence exerted by the representatives of dialectics in their criticism of traditional formal logic with its principle 'either it is the case, or it is not'. But many-valued logic took the course of a generalization of two-valued logic, retaining all the necessary foundations of constructing logic in general.

Notwithstanding its more refined approach to the evaluation of human knowledge and to the laws of thinking, many-valued logic does not turn into dialectics or into one of its divisions, and in no way confirms any of the principles of dialectics. All attempts to make use of the idea of many-valued logic in an opposite sense can only lead to confusion and fruitless terminological disputes.

§ 7. Besides cases of the type of Zeno's 'paradox', other examples which seem to diverge from logic can also be given, but as a matter of fact they again create this illusion by confusing concepts, obscuring them, ignoring necessary distinctions, etc.

There are propositions whose structure can be described by the following schema: $x = Df.\ K^*yz$. The symbol K^* is here a special sort of ordered conjunction: K^*yz is true, if y is true and z is true; K^*yz is false, if y is true and z is false; K^*yz is undetermined, if y is false. Similarly x is such a function of y and z. This type of function is quite conceivable, although they are normally not used. An example for their interpretation will be given below. Two kinds of negations of x are possible:

1) N^1x means that x is false;

2) N^2x means that x is undetermined (does not appear to be true).

Also possible is the negation N^3x, meaning: x is not true (but it is unknown whether it is false or undetermined).

Take the proposition 'Peter gave up smoking'. It is equal in meaning and equivalent in truth value with the ordered conjunction of the two propositions 'Peter did smoke (up to such-and-such a time)' and 'Peter does not smoke (now, since such-and-such a time)'. Its negation 'Peter did not give up smoking' can mean 'Peter did smoke and Peter still smokes now' and 'Peter never did smoke'; it can also mean the negation of the truth of the original proposition without assigning falsity or indeterminacy (leaving this unknown). Ignoring such a hidden structure of the proposition and of the distinctions indicated gives rise to the well-known 'paradox' of the question 'Have you stopped beating your wife?', where, in the case of accepting as truth values only truth and falsity, both 'yes' and 'no' amount to a confession that the person answering did beat his wife up to a certain time.

§ 8. Let us make some remarks in addition to what has been said above (in ZINOV'EV 1961a, 1962 the corresponding considerations are set forth in more detail).

If the unity of science is accepted not as something which is static and established once for all, but as a tendency of essential importance inherent in the changeable variety of connections and relations of ideas, theories, problems, directions, etc., then logic as a whole can be spoken of as a completely real fact. The question arises: from the point of view of the number of truth values, of which kind is logic as a whole? Obviously, one must admit that logic as a whole is neither two-valued nor many-valued. And this not only and also not so much because many-valued logical systems did arise, on whose basis a special trend has arisen in logic, but first of all, because the question of the number of truth values never has been and never will be an initial question in logic understood as a special sphere of scientific investigation. It is an initial question only in the construction of individual logical theories, each of which taken separately does not exhaust the content and the problems of the science taken as a whole.

The principles of constructing individual logical systems and those of logic as a whole are, naturally, different. Among these differences it is important to note the following. In considering logic taken as a whole, as a special branch of scientific investigations, we discover that in its initial foundations, considering the source of its problems and the criteria for evaluating its results, it is an empirical science. This seems to be already shown by the fact that the immediate empirical reality with which it is concerned, first of all, is language. In any case, considerations occur in logic either obviously or covertly which are empirical in character and form its foundations. In the development of logic in the last hundred years, which has been preoccupied largely with working out logical calculi and the general principles of their construction, these empirical foundations of logic have been relegated to the background, but they have not been eliminated: however logic may be understood, it is discernible by the attempts to analyze its general concepts 'proposition', 'predicate', 'true', 'follows', etc. Numerous logical systems are constructed with the direct purpose of making precise and systematically developing the properties of such ordinary phenomena of language as the symbols 'and', 'or', 'if... then...', 'necessary', 'possible', 'should', 'may', etc. As for individual logical systems, in the majority of cases their construction is decisively determined by the task of making out the conditions of deduction from certain assumed presuppositions. Thus it is possible that no attention is paid to what would be the empirical considerations. Many cases are known where some theoretical construction can only be called logical if its connection with numerous other investigations is taken into account (e.g. the formal construction of many-valued propositional logic by Post).

In the case under consideration, reference to the empirical foundations of logic is of considerable importance. Indeed, speaking of two-valued or many-valued logic, we are forced to speak of propositions and of truth values of propositions. But what are these? In constructing individual logical systems this question can be circumvented in different ways:

1) the corresponding concepts can be accepted as undefined or intuitively clear;

2) the symbols of propositions and of truth values can be considered as symbols which can be interpreted in terms of extremely different object domains (which can be translated in the language of the latter);

3) or in general, a more general terminology of 'argument', 'function', 'value of the argument', etc. can be used.

But this is no solution of the problem within the limits of logic as a whole.

Obviously, the answer to the problem stated should first of all be included in the description of the process of abstracting propositions as determinate symbolic structures based on the observation, selection, analysis, comparison, modification and standardization of the way in which the immediately given facts of language are represented and in which we are accustomed to operating with them. It is sufficient to note that the subject-predicate structure of propositions (propositions with many-place predicates included) can in predicate logic be taken as something self-explaining or as something postulated only so far as in logic it already has been abstracted from the empirically given and as the abstraction itself has been so clearly justified that it can be understood even by a schoolboy.

But the structure of propositions does not depend on the number of truth values, and one cannot speak of the definition of the latter until the propositions themselves have been abstracted, given a standardized expression in logic (it might be recalled that we are speaking here of the order of succession in the construction of logic as a whole). The truth values are properties of propositions. Naturally, the concept of proposition has to be first introduced. To these general considerations in the given case one adds a specific one concerning the truth values: in the language observed they cannot be distinguished on the same basis as the propositions themselves. The propositions are symbolic structures. In a great many cases, chosen as examples for the existence of their abstraction, they coincide with the visible or audible articulation of statements (in any case, the form of the statements can be given a uniform appearance). The truth values, however, are not elements of the structure of the language. They are connections of correspondence between propositions and situations which are different from them, and they are the classified results of a confrontation between the first and the latter. This is shown by their customary interpretation, which in science can be represented by the relation of two sorts of symbols: symbols of propositions and symbols of objects, properties, relations, connections, etc., with which the propositions are confronted. Of course, in order to determine the types of confrontation of propositions and object situations different from them it is necessary to have some previous knowledge about the structure of the propositions themselves. It should be stressed that here as well as below, we are not speaking about the empirical verification of propositions, but about the definitions of truth values which in particular can be employed in carrying out such a verification.

Schematically, the meaning of the expression 'truth value i' can be explained as follows: the expression 'proposition x has the truth value i' means that the confrontation of the proposition x with some given situation gives such-and-such a result; more exactly, that this given situation is described or can in principle be described by the proposition y (y can be identical with x). This schema is the schema of the construction of definitions of truth values. As a particular case of a realization of it one can mention Tarski's schema: the expression 'proposition x is true' means that a given situation, with which x is confronted, is described by this very proposition x.

Of course, the schema given here does not eliminate the many difficulties connected with the definition of truth values. But it is quite sufficient for explaining the fact that the truth value of a proposition is not an element of its structure but a type of relation between it and some situation (a real one or a hypothetical or assumed one). Thus we must obviously take into account the structure of the propositions if we want to pass from the general schema to its realization for two or more truth values. Simply admitting some finite (or infinite) number of truth values in no way means that each individual one of them is defined according to the given schema.

Even the most simple proposition contains two descriptive symbols: subject and predicate. And this means that there are three possible variants in confronting it with some situation:

1) in the given situation there is no object corresponding to the subject;

2) such an object exists but it does not have the property represented by the predicate;

3) such an object exists and it does have the property represented by the predicate.

If an even more complex structure of propositions must be taken into account, then the number of possible variants of confrontation of the propositions and the situations increases. Thus the number of truth values is not limited by some absolute natural necessity. It is limited by historical circumstances, and, in the theories reflecting them, by the approach used in classifying them.

The limitation of truth values to two in number has been linked in the history of logic with very different motives. We only mention two of them. First of all, in logic, the cases of the type described in the first variant above have been tacitly excluded. In a more general sense: in logic, only those cases were taken into account where the evaluation of the proposition was dependent on one and only one verification step (in the example: examining whether the object has the property or not). Secondly, the propositions were evaluated according to the principle: either the state of affairs is actually such as it is described in the proposition or it is not (it is different). Thus, the meaning of 'it is not' (negation) was made clear by the context and was in general not analyzed.

Still, a whole series of reasons can be given why the number of truth values is in principle unlimited. For instance, because of the ordering of the propositions (or, correspondingly, of the events described by them), an example of which has been given in the preceding section; because of the community of subjects and predicates (cf. the following chapter); because of the modalities, the degrees of probability, etc.

Thus, if the truth values are understood as relations to certain facts, events, objects, etc., then it is clear that any number of truth values can be introduced: these relations are set up by man, and there is in the nature of the propositions and the states of affairs, with which the propositions are confronted, no absolute necessity of limiting oneself to two types of relation. But this is only an abstract, conceivable possibility which need not necessarily be realized. It is not obligatory to introduce three or more truth values, if this is not dictated by important reasons. If some human needs of knowledge are satisfiable by two truth values, then there is, naturally, no reason to assume the additional burden of considering more than the two truth values or to complicate our reasoning. But if the point of view of two-valuedness constitutes an obstacle for the solution of some problems, and the point of view of many-valuedness offers some advantages, then from an epistemological point of view there are no natural reasons which could prevent man from introducing into the number of his tools of knowledge many-valued constructions and a many-valued conception of logic.

It might be stressed that we do not present man with the alternative: either two-valuedness or many-valuedness. There is another alternative: either human knowledge is evaluated only as true or false, or, other evaluations besides truth and falsity are possible. In the second case, truth and falsity retain their meaning. More than that, they play here a more important rôle than just any two truth values among others: the evaluation of knowledge as true and false is historically and logically basic, simpler and more often used; in constructing or interpreting many-valued systems two values are selected which are understood as truth and falsity (or as corresponding to truth and falsity); in defining the tautologies (the true formulas, the always asserted propositions) the set of truth values is divided into two subsets where one corresponds to truth and the other to falsity, etc. We have remarked above that in a well-known sense two-valuedness is a presupposition of all reasoning.

§ 9. The attempts to consider the logical structure of concrete sciences (of quantum mechanics in the first place) from the point of view of many-valued logic led to the formulation of the problem of the universality (or more exactly: of the non-universality) of logic.

Whereas we understood by the problem of the plurality and unity of logic the mutual relations of the logical theories within the science of logic irrespective of their specific applications, by the problem of universality we mean something else. Logical theories can not only be interpreted in terms of some concrete object domain (they can not only be considered as a formal model of this domain, as an apparatus for solving its particular problems), but they can also be considered as theories of thinking, as descriptions of the rules of scientific thinking. Are the rules of thinking described by these theories universally valid or not?

A negative answer to this question seems to suggest itself: we saw above that in the domain of mathematics a limitation is placed on the law of the excluded middle; in the domain of microphysics facts concerning the indeterminateness of propositions must be taken into account; etc. However, because of the ambiguity of the very concepts of universality or non-universality, the categoricity of judgments on the given question are doubtful.

We can reason in the following way: logic reflects some properties of reality and is based on ontological generalizations; consequently a difference of the properties of different domains of reality (e.g. of microcosmos and macrocosmos, of finite and infinite sets) leads to a difference between the systems of logical rules for these domains (cf., e.g., DESTOUCHES-FÉVRIER, DESTOUCHES). But the reasoning can also be constructed in another way: regardless of the differences between the properties of different spheres of reality, there are some general properties reflecting the rules of logic: it would be absurd to thus divide the world up into spheres which need different rules of logic for their reflection.

Furthermore, take this example: in some reasoning, say, a conclusion is drawn according to the schema of *modus ponens*, whereas in another the schema of the first figure of the syllogism is used; the question is whether these schemata are universal. Clearly, the very concept of universality is here ambiguous, and this excludes a simple 'yes' or 'no' answer. The presence of descriptions of different rules of thinking in one and the same logical theory shows that logic is not universal in the following sense: different conditions of knowledge can require different logical rules. However, under standard conditions, standard rules have to be applied, and in this sense of the word, logic is universal.

Many-valued logic is not an exception on this account.

Finally, from the material considered here it follows that the laws of two-valued logic are not denied by the laws of many-valued logic, nor are the laws of the second denied by the laws of the first. Therefore, it would be a grave mistake to interpret the non-universality of the laws of logic as a division of the world into spheres some of which are subject to certain logical laws, whereas in others their negation holds. The non-universality of the laws of logic must obviously be understood as taking into account the factual mutual relations between the logical systems describing them. But to answer the question of how this appears in scientific investigation presupposes a concrete analysis of the content and the history of those sciences where attempts are made to use the ideas and the apparatus of many-valued logic. Within the limits of logic and the general philosophical setting according to which the rules of logic ultimately reflect some general properties of the things and the historically developing objective conditions of their knowledge, only the following can be said: the non-universality of the rules (laws) of logic expresses the differences in the orientation of investigation, the wider or narrower comprehension of the objects of reality, the greater or smaller differentiation in evaluating the relation between propositions and reality, penetration into the more complex relations and connections between the objects, etc. As to the rise of many-valued logic, it aims (among other problems) at giving a wider and more exact description of the laws of scientific thinking, and at accounting also for the progress of science in this respect, but it in no way denies the laws of classical logic. It denies only the claim of some representatives of traditional logic for a final completeness of formal logic and for the independence of its laws from the progress of knowledge.

Some words on the evaluation of scientific theories from the point of view of the character of logic: every scientific theory is unambiguous in the sense that all its propositions are intended to be true and the rules of definition and deduction appear as algorithms for the construction of new laws and propositions out of the ones already given. From this point of view the logical theories are unambiguous: two-valuedness and many-valuedness belong here to the content of the theories, and are properties of the propositions considered in the logical theories.

Every scientific theory is two-valued or in general n-valued in the sense that the logical symbols 'and', 'or', 'if... then...', etc., are used in it, and the rules of deduction can be defined and justified in logical theories in which two or generally n truth values are attributed to the propositions.

Thus the choice of one or another system of definitions depends on the properties of the object domain studied and on the conditions of its knowledge. Thus, if in the science, given propositions can be formed which under the given conditions cannot be proved or refuted, then it is completely natural to introduce a third value, and also to take this circumstance into account in the formulation of the logical rules. As to the choice of the 'best' variant, this depends on the concrete conditions of the science and on how adequately they are reflected by the one or the other logical theory.

CHAPTER SIX

GENERALIZED QUANTIFIERS

§ 1. The ideas of many-valued logic cover nowadays not only the domain of propositional logic but also the domain of the logic of predicates (we are thinking here mainly of the work of Rosser and Turquette). But whereas in propositional logic many attempts have been made to give an informal interpretation to many-valued constructions and to make use of them in the solution of concrete scientific problems, the same is not the case with the many-valued logic of predicates: up to now no need has been recognized for quantifiers different from the classical existential and universal quantifiers, and no serious attempt whatsoever has been made to give an informal interpretation of the many-valued logic of predicates, even if it is only with the aim of illustrating the very possibility of such an interpretation.

But if this is the situation, then perhaps it is not worthwhile to discuss this subject? It seems that this is not the case. In science the proposal of theoretical results itself often generates the demand for them. The development of logic and of its applications in the last decades has shown that logic is not only able to describe customary ways of reasoning, but also to propose new ones which have not yet been incorporated into the number of logical tools of science. One cannot rule out the possibility that this will also be the case with the many-valued logic of predicates. And one also cannot rule out the possibility that the analysis of the fundamental ideas lying at their basis will in some measure promote the clarification of their meaning and stimulate attempts for making concrete use of them. Below we only will treat briefly the very basic questions concerning the possibility of many-valued quantifiers. Before that we will consider some general questions of the logic of predicates.

§ 2. In propositional logic the structure of propositions is considered only insofar as it is covered by truth functions: complex propositions are analyzed (ultimately) into atomic propositions connected by determinate symbols into a whole. The atomic propositions are taken as elementary and are not analyzed into parts which are no longer propositions. This holds for two-valued as well as for many-valued systems.

In the predicate calculus this limitation is somewhat loosened: the subject-predicate structure of the propositions is here taken into account. We have said 'somewhat loosened' since the logic of predicates does not exhaust the problems of the construction of propositions actually used in ordinary and scientific speech.

But however this may be, in predicate logic the structure of propositions is taken into account, so that here we are dealing with logic in the age-old meaning of the word. And as predicate logic is based on and includes the logic of propositions, any doubt concerning the logical character of propositional logic in general and many-valued propositional logic in particular should no longer arise.

We will start here from the empirical fact that every proposition can be analyzed into two parts, and that the reader is accustomed to such an analysis. In one of these parts symbols (terms) designating objects (i.e., that about which the proposition speaks) are included, and in the other of these parts, symbols designating properties of the objects (i.e., that what is said of the objects, what is attributed to them). The first are called 'subjects', the second 'predicates'. The confusion of objects and properties on the one hand and subjects and predicates on the other hand, which can be found in many works, does not lead to entanglement (as, e.g., in HILBERT and ACKERMANN) except as far as a correspondence between them is presupposed.

The analysis indicated is not always the result of immediate observation, as in the case of propositions of the type 'number 29 is an integer', 'the electron is negatively charged', etc. It may be the result of a sufficiently high degree of abstraction. For instance, in the proposition 'number *5* is larger than number *3*, whereas number *2* is smaller than number *3*', the subjects 'number *5*', 'number *3*', and 'number *2*' are separated in space by other words and the subject 'number *3*' occurs twice. But we count them as one part and pay no attention to the repetition. The predicate of this proposition, corresponding tot he enunciated property of the numbers, taken together, cannot generally be delimited in space from the subjects without the further abstraction which we explain below.

The propositions also contain an indication of whether or not the property belongs to the subject: a symbol of attributivity. This is again an empirical fact, a determinate custom. It can be expressed in different ways: by the order of the words, by their grammatical form, by special words. We will represent it by writing subject and predicate in a determinate order together, as this is usual in the contemporary logical literature. The subjects we will designate by small Latin letters $a, b, c, \ldots$, the predicates by large Latin letters $P, Q, R, \ldots$ The propositions we will formulate by symbols of the form $P(a), Q(a), P(a, b), Q(a, b, c), NP(a)$, $NQ(a), NP(a, b), NQ(a, b, c)$, etc. These symbols are to be read: 'object *a* has the property *P*' (or simply: '*a* has *P*', '*a* is characterized by the fact that *P*'), '*a* and *b* are characterized by the fact that *P*', '*a* does not have *P*', etc. The expressions of the form 'object *a*', '*a*', 'property *P*', '*P*' mean correspondingly 'the object designated by the symbol *a*' ('the object called *a*') and 'the property designated by the symbol *P*'. The symbol *N* means here the negation of the fact that the properties belong to the objects.

Attributivity is not expressed by a special symbol beside the symbols of subjects and predicates simply because in predicate logic it has a single function, to ratify the relative order of subjects and predicates. Thus the articulation of the propositions into two parts has no relation to the question of the number of parts of the propositions asked irrespective of the purpose of analyzing the propositions.

The understanding of the given propositional notation for propositions with one-place predicates (predicates with one subject) presents no difficulty, since it coincides with the structure of propositions of a type often met: 'number *11* is prime', 'the window is open', etc. As to propositions with predicates of two or more places (with two or more subjects), a further abstraction must be made for the analysis of the predicates. This abstraction consists in the transformation of the propositions of the form 'number *a* is larger than number *b*', 'point *a* does not lie between the points *b* and *c*', 'phenomenon *a* is the cause of phenomenon *b*', etc., into, respectively, 'number *a* and *b* are characterized by the fact that the first (in the order of writing) is larger than the second', 'the points *a*, *b* and *c* have the property that the first does not lie between the second and the third', 'the phenomena *a* and *b* are such that the first is the cause of the second', etc. (cf. ZINOV'EV 1959c). This abstraction, which (by convention) does not affect the content of the propositions, permits the covering (from a determinate point of view, of course) of all possible propositions so that this way of abstracting this kind of structure from the empirical facts of speech can be accepted as a way of introducing the very concept of proposition.

The way indicated for distinguishing the predicates is completely equivalent to the way set forth in HILBERT AND ACKERMANN: *P( )*, *P(,)*, *Q(, ,)*, etc., where the empty places can be filled by subjects. But in such a notation the order of the subjects, which plays an important rôle, is not taken into account, so that in every case a convention must be made concerning it.

Up to now we have not distinguished between properties and relations: the latter are, in the terminology adopted, a particular case of properties designated by predicates of several places. Thus it should be stressed that relations are not simply designated by words like 'larger', 'smaller', 'cause', 'between', etc., but by expressions of the form 'such-and-such in order is (are) larger than such-and-such in order', 'such-and-such in order are the causes of such-and-such in order', etc. There is also another way possible: by properties and relations one may understand respectively what is designated by predicates with one or several places. But this way has no advantage over the first one, so that their difference is purely verbal.

In some works (e.g. in ROSSER AND TURQUETTE) the propositions *P(a)*, *Q(a, b)*, etc. are counted as predicates. But in our terminology the predicates as well as the symbols of objects are only parts of the propositions, and not propositions themselves (such a terminology is adopted in the majority of cases).

§ 3. Subjects and predicates differ in degree of generality. With respect to the subjects this is obvious ('object' – 'number' – 'real number'–'integer' – '*5*'). As for the predicates, we can show this by the following example: a particular case of the property of being changed is the property of being diminished, and a particular case of the latter is the property of being diminished by two; the predicates designating them have correspondingly smaller degrees of generality.

The degree of generality does not prevent any symbol from becoming a subject or a predicate. Insofar as we understand a proposition as being some symbolic structure, expressions of the type 'the object has the property', 'the object does not have the property', 'two objects are characterized by the property (stand in the relation)', etc. are also propositions. Semantical considerations are here not essential, since for the explanation of the questions which interest us we can always assume for the sake of simplicity that the symbols of the properties and objects are not empty with respect to their designation.

The distinction of the subjects and predicates according to their degree of generality is in predicate logic generalized by the concepts of constant and variable subjects and predicates.[11]) These concepts carry with them additional difficulties which we will treat below. For the moment we will make two remarks on what has been said. In logical literature – often because of the terminological confusion between objects and properties on the one hand and subjects and predicates on the other hand – there is a corresponding confusion with regard to variables and constants. We will relate the latter characteristics exclusively to subjects and predicates. In fact, in place of a subject variable, e.g., a subject constant can be substituted, but in the place of a table in general, e.g., no table related to some narrower class and no individual table can be substituted, since the first does not at all exist alongside the others. The second remark concerns the interpretation of the subject constants as symbols of individual objects, an interpretation widely accepted. This interpretation is clearly inexact since general symbols also can occur as constants. For instance, if as object domain the domain of numbers is given, the word 'number' will play the rôle of a variable in a proposition like 'the number is not prime'; as a constant, not only the symbol of a concrete number but also the symbol of a class of numbers can occur; in particular, if we take as a constant the expression 'the number x^n (where x and n are integers larger than *1*)', then by substituting it for the word 'number' we get a definite (in the terminology of Hilbert) proposition, namely a true one.

Take the proposition 'number *a* is prime' where the expression 'number *a*' refers to certain numbers. Is this proposition true or false? Of course, it all depends on what number we will confront this proposition with (or the symbol of what number we put in the place of the expression 'number *a*'). If it is one of the numbers *11* and *13* then the proposition will be evaluated as true, and if it is one of the numbers 8 and 9, as false. In this way the admission of general symbols permits the construction of propositions which will be true if referred to some objects and properties and false if referred to others (in the case where there are two truth values). The substitution of subject and predicate constants in the place of variables also gives a definite truth value to such a type of proposition.

Subject and predicate variables and constants can, generally speaking, also be understood thus: the variables are symbols of objects and properties of such a degree of generality that the content of their propositions has different truth values depending on what objects and properties they are confronted with; the constants are those symbols of objects and properties whose substitution in place of the variables gives a definite truth value to the propositions. A proposition containing subject and predicate variables is undetermined in the sense that its truth value can change depending on the situation with which it is confronted. The substitution of subject and predicate constants in place of the variables (a substitution in place of all variables is presupposed) expresses this confrontation also, since it is by means of the constants that this factual situation is described.

If proposition contains some variables, then it can only be changed into a definite proposition if constants are substituted in place of the variables. In this way the distinction between constants and variables appears for the concrete cases of propositions to be completely relative: one and the same subject (and predicate) can for one combination of substitutions appear to be a constant, and for another a variable. For example, if in the proposition 'the integer *a* divided by the integer *b* leaves no remainder' *1* is substituted in place of *b*, then we get a definite proposition and *a* will be counted as a constant; if, however, number *2* is substituted in place of *b*, then *a* becomes a variable. Therefore the definition given of subject and predicate constants and variables gives no absolute criterion for their distinction in concrete propositions. Here only the following can be said: if in a proposition the substitution of any subject (or predicate) whatsoever for a given one gives a definite proposition, then the subject (or predicate) given is the only variable; if the proposition becomes undetermined, then either it contains other variables or the subject (predicate) substituted is itself a variable. Thus, logic does not fix some single interpretation of some given proposition as the only possible one, because it depends on the context as to which interpretation is chosen.

The expressions containing variables can be considered as schemata or forms of propositions (cf., e.g., HILBERT and ACKERMANN, ROSSER and TURQUETTE) which only through substitution of constants are transformed into propositions. In this case the definition of 'proposition' by means of 'truth value' (in two-valued logic it is assumed that it can be either true or false) is tacitly presupposed. This way involves, it seems to us, a vicious circle: the definition of the terms 'true', 'false', etc. presupposes that one has already the concept of proposition as of a determinate symbolic structure (if, namely, one intends a definition of terms). Moreover, ordinary and scientific speech are full of propositions with subject and predicate variables which contain no quantifiers, and to exclude them from the number of propositions is simply not sensible. Finally, it can be shown that the systems of terms introduced in these two different ways appear to be isomorphic: the proposition forms correspond to undetermined (in the sense indicated) propositions, the propositions to the definite propositions.

What has been said above covers not only the cases of two-valued propositions, but also generally the cases with any number of truth values. In this way the undetermined propositions can be considered as n-valued in the full sense of this word. Actually this is as yet only an abstract, conceivable possibility for obtaining, as a result of substituting different subject and predicate constants in place of the variables, n types of propositions differing in their truth value. However, an informal illustration of this can be given.

The set of subject and predicate constants, whose substitution in place of the variables gives definite truth values to the propositions, we call the 'range of values' of the subject and predicate variables, and every one of these constants we call a 'value' of a corresponding variable.[12]) We shall consider propositions with a one-place predicate. Let P be a predicate constant and a a subject variable. Then the following is possible:

1) $P(a) = 1$ for some values of a;
2) $P(a) = 2$ for other values of a;

$\vdots$

n) $P(a) = n$ for a nth group of values of a.

In general, for some set of values of a from all its possible values, $P(a) = i\ (1 \leqslant i \leqslant n)$. Similarly for a variable P, for variable P and a, for two and more subject variables. The case where a complex proposition contains several different kinds of predicate variables (predicates of the second, third, etc., order; cf. HILBERT and ACKERMANN), we do not consider. However, what has been said also has reference to it in a general form, since also in that case the truth values of the propositions can be denoted by a system of equations of values.

In traditional logic, variables and constants also occurred in an implicit form. The axiom 'the characteristic of a characteristic is a characteristic of the thing' expresses in some sense the fact of substitution, if it is given the following form: 'if it is true that $P(a)$ and if b is a value of a, then it is true that $P(b)$'. For predicates, however, such an axiom cannot be accepted. Thus, if as values of the predicate of the proposition 'the temperature of a given mass of gas increased' symbols of quantitative degrees of increase (e.g. 'increased twice') are accepted, then from the truth of this proposition it cannot be inferred 'the temperature of a given mass of gas increased twice'. The inference, of course, is only valid in the opposite direction. Thus, if $Q(a)$ is true and Q is a value of P, then it is also true that $P(a)$. However, a terminological uniformity can be established here which makes the definition of the value of a predicate variable conform to the definition of the value of a subject variable. Along the line considered, i.e. along the line of passing from the general to the particular (and vice versa) the many-valued evaluation of propositions does not touch traditional logic, which remains an independent fragment of logic.

The many-valued evaluation of propositions in no way touches the structure of the propositions. From the point of view of the content of this section it speaks only about the differences in the mutual relations between propositions with subject and predicate variables, and situations with which they are confronted, which is also expressed in the classification of the values of the variables. We give two simple examples. Take the proposition 'number *4* has the property *P*'. Suppose the range of values of *P* is not limited. Then three classes of values are possible:

1) values of *P* which transform the given proposition in a true one ('number *4* is an integer', 'number *4* is even', etc.);

2) values of *P* which transform this proposition into a false one ('number *4* is irrational', 'number *4* is odd', etc.);

3) values of *P* which transform this proposition into a meaningless one ('number *4* is yellow', 'number *4* is bourgeois', etc.). Meaninglessness can here be considered as a third truth value.

Take, further, the proposition 'object *a* has a positive charge'. Again there are exactly three possible classes of values of *a*:

1) transforming the proposition into a true one ('proton');

2) transforming the proposition into a false one ('electron');

3) transforming the proposition into an undetermined, meaningless, not verifiable and not falsifiable one ('centaur').

We gave, so to speak, examples from 'life'. But such cases can also be found in science (cf., e.g., BOČVAR 1938).

§ 4. There is another way of giving definite truth values to propositions containing variables. It consists in binding the variables by means of quantifiers, i.e., by means of special signs indicating the number of objects and properties covered by the subjects and predicats. It is not literally a question of counting but only of taking into account the quantitative aspect in a very rough way, namely by means of the universal ('all', 'for all') and the existential ('there exists', 'some', 'for some') quantifiers designated (in particular) by the symbols Π and Σ. Their meaning is supposed to be clear from their customary use. Their negation $N\Pi$ and $N\Sigma$ signifies correspondingly 'not all' and 'there does not exist' ('no one', 'for no one'). The propositions take the forms: $(\Pi a)P(a)$, $(\Pi P)P(a)$, $(N\Pi a)(\Sigma b)P(a,b)$, etc., which are to be read: 'all a have P', 'for all P, a has P', 'not for all a, there is a b such that $P(a, b)$', etc. Accepting Π and $N\Pi$ as basic, Σ and $N\Sigma$ can be defined: $(\Sigma a)P(\alpha) = (N\Pi a)\,NP(\alpha)$, $(N\Sigma a)P(\alpha) = (\Pi a)NP(\alpha)$; similarly for ΣP and $N\Sigma P$; α contains a or is identical with a. From the definitions, it follows that $NN\Pi = \Pi$ and $NN\Sigma = \Sigma$.[13])

Propositions containing subject and predicate variables are transformed into definite propositions, if either all variables are replaced by constants, or all are bound by quantifiers, or one part of the variables is bound by quantifiers and the other replaced by constants. Only one thing is important: every one of the quantifiers here employed binds only one subject of predicate variable, and so we call them one-place quantifiers. For example, the proposition 'number a is even' contains a single subject variable 'number a'. This variable can be bound by the quantifiers Π and Σ and we obtain the proposition (Πa)(a is even), (Σa)(a is even). The first is false, since not every number is even, the second is true, since there are even numbers. The proposition 'an even number has the property P' contains a single predicate variable 'property P'. Binding it with quantifiers we get: (ΠP)(an even number has the property P), (ΣP)(an even number has the property P). The first is false, as there are P which do not belong to even numbers, e.g., the property of not being divisible by two. The second is true, since there are such P, e.g., the property of being divisible by two. The proposition with two variables 'number a is divisible by number b without remainder' can with the help of our one-place quantifiers be transformed into a definite proposition in exactly four ways: $(\Pi a)(\Pi b)\,Q(a,b)$, $(\Pi a)(\Sigma b)Q(a,b)$, $(\Sigma a)(\Pi b)Q(a,b)$, $(\Sigma a)(\Sigma b)Q(a,b)$

(where Q is the predicate of the proposition). Only the fourth proposition is true, all the others are false.

The law of the excluded middle has, for the propositions $P(a)$, the following form: $(\Pi a)(\Pi P)\{$either $P(a)$ or $NP(a)\}$ or, making use of the concepts of the calculus of propositions $(\Pi a)(\Pi P)\{A[P(a)][NP(a)]\}$, i.e., it contains two quantifiers and the predicate A (similarly the law of contradiction can be formulated taking into account the structure of the propositions).

In traditional logic the quantifiers have still another function, namely they are used to express the relations between general and singular propositions. Thus, for example, the following laws are well known in traditional logic:

1) if $(\Pi a)P(a)$ holds, then $(\Sigma a)P(a)$ holds also;
2) if $(\Pi a)NP(a)$ holds, then $(\Sigma a)NP(a)$ holds also;
3) if $(\Sigma a)P(a)$ holds, then $(\Pi a)NP(a)$ does not hold;
4) if $(\Sigma a)NP(a)$ holds, then $(\Pi a)P(a)$ does not hold;
5) if b is a value of a and $(\Pi a)P(a)$ holds, then $P(b)$ holds also.

In predicate logic they are sometimes covered by the rules of substitution, sometimes by other rules of derivation, sometimes by the definitions of the quantifiers and the assertions derivable from them, and sometimes by assertions derivable from axioms. For example, if the axioms

1) $C\{(\Pi a)P(a)\}\{P(b)\}$,
2) $C\{P(b)\}\{(\Sigma a)P(a)\}$,

(cf. HILBERT and ACKERMANN) are accepted, then by the rules of deduction we get: if axiom (1) holds and $(\Pi a)P(a)$ holds, then $P(b)$ holds; if axiom (2) holds and $P(b)$ holds, then $(\Sigma a)P(a)$ holds; i.e., $C\{(\Pi a)P(a)\}\{(\Sigma a)P(a)\}$, the first of the assertions given above, holds. Similarly (or by using double negation) also the second of the given assertions of traditional logic can be obtained. In general, the principles of traditional logic connected with the relation between the general and the singular (and the particular) are in one way or an other covered by predicate logic. And, consequently, the problem is reduced to the following one: do the one-place quantifiers exhaust all possibilities of quantification or not?

Every one-place quantifier binds only one subject or predicate variable. Obviously, according to the very meaning of the quantifiers the two-membered relations 'either all..., or not all...' and 'either there exists..., or there does not exist...' (in the sense 'either some..., or no one...') are fulfilled. Suppose x is a proposition all of whose variables with the exception of a single variable α, are bound by quantifiers or replaced by constants, so that the transformation of x into a definite proposition depends exclusively on α. Then what has been said above can also be written in the form: $A\{(\Pi\alpha)x\}\,\{(N\Pi\alpha)x\}$, $A\{(\Sigma\alpha)x\}\,\{(N\Sigma\alpha)x\}$. Thus the binding of α by any of the quantifiers transforms the proposition either into a true or into a false proposition. And in this (and only in this) sense the quantifiers Π and Σ are two-valued.

In general, the quantifiers which transform propositions into two-valued propositions (propositions with all variables bound being either true or false) we will call 'two-valued'. One-place two-valued quantifiers we will call 'classical'.

The fact that only two classical quantifiers are used is especially connected with the two-valuedness of logic, more exactly, with the two-valuedness of negation:

$(\Sigma\alpha)x$ is simply another form of $(N\Pi\alpha)Nx$, or $(\Pi\alpha)x$ is simply another form of $(N\Sigma\alpha)Nx$. The question as to which one of the two quantifiers has to be taken as primary is an interesting epistemological question, but it goes beyond the limits of our subject. We note here only that in principle the introduction of other one-place quantifiers is possible, e.g. the following:

1) 'anyone whatsoever' ('anyone whatsoever of the customers (but not more than one) could buy this thing');

2) 'one and only one' ('one and only one of these people is able to fulfil this task').

It may be stressed that this introduction is possible on purely formal grounds. It can also easily be noted that the negation of these quantifiers is not unambiguous.

§ 5. The question of the introduction of quantifiers with respect to which the classical quantifiers are only a particular case, and also the question of the axiomatization of a predicate logic with such quantifiers, has in a more or less detailed way, been considered in ROSSER and TURQUETTE. Here we consider only one question, namely the question of the introduction of non-classical or generalized quantifiers (we will make use of the ideas of ROSSER and TURQUETTE only) i.e. of quantifiers which can bind more than one variable (many-place quantifiers) and can transform propositions into many-valued propositions (many-valued quantifiers). The interest in the many-place quantifiers is here caused by the fact that the problems connected with them are closely entwined with the problems of many-valued quantifiers. Strictly speaking, without the first the second would be fairly poor in content.

The method of generalizing is based on the following ideas:

1) a proposition can take more than two truth values depending on the values of the subject and predicate variables;

2) a construction of quantifiers which bind simultaneously more than one variable is possible.

In ROSSER and TURQUETTE the assumption of functions of the form $F^i(a^1,\ldots,a^m, x^1,\ldots,x^k)$, where $m \geqslant 1$ and $k \geqslant 1$, the a's are subject variables, the x's are propositions, is primary. Particular cases of such functions are functions of two-valued logic: $F^1(a, x) = (\Pi a)x$, $F^2(a, x) = (\Sigma a)x$, where a is a subject variable (at least one of the subjects) of x. As is noted in ROSSER and TURQUETTE, nobody has found quantifiers with $m>1$ or $k>1$ necessary, but they are possible.

Consider the following example. Suppose the following conditions are given:

a) a and b are subject variables;

b) $x = P(a, b, c^1,\ldots, c^m)$, $y = Q(a, b, c^1,\ldots, c^m)$;

c) x and y take different truth values from 1 to n dependent on the values of the variables a, b, $c^1,\ldots, c^m$; the truth values of x and y may depend on only a part of them;

d) a proposition of the form $F(a, b, x, y)$ is considered. We determine the truth value of this proposition by the following prescriptions:

1) $F(a, b, x, y) = 1$ for a given set of values of $c^1,\ldots, c^m$, if and only if there is such a value of a that for the given values of $c^1,\ldots, c^m$ and for all values of b (from a given object domain) $Cxy = 2$;

2) $F(a, b, x, y) = 2$ for a given set of values of $c^1,\ldots, c^m$, if and only if for every value of a and b from the given object domain and for the given set of values of $c^1,\ldots, c^m$ $Kxy = 1$;

3) $F(a, b, x, y) = 3$ for a given set of values of $c^1,\ldots, c^m$, if and only if for every value of a and for the given values of $c^1,\ldots, c^m$ there is a value of b from the given object domain such that $Cxy = 3$;

4) $F(a, b, x, y) = 4$ for the values of $c^1, \ldots, c^m$ in all the cases except in those indicated in the above points;

5) $F(a, b, x, y)$ does not take the values $5, \ldots, n$.

The symbol F defined in this way has all the properties of a quantifier: it takes into account the number of values of the variables and binds the latter. But it possesses several peculiarities in comparison with the classical Π and Σ:

1) it binds two or more variables;

2) it transforms propositions into n-valued propositions (i.e., into propositions taking one of n truth values, where $n > 2$);

3) it is defined by means of the classical quantifiers which each bind only a single variable; so that a contradiction between many-valued and two-valued logic is impossible also along these lines.

In introducing quantifiers like F, symbols of two-valued propositional logic (in addition to the classical quantifiers) may be used, so that also for the construction of many-valued predicate logics, two-valued logic is sufficient. We give an example for the case $n = 4$. Let xi signify that $x = i$ $(1 \leqslant i \leqslant 4)$. Then $F(a, b, x, y)$ is defined by the following statements (every line is a partial normal form, i.e. it shows, when the propositions take the truth value equal to the number of the line):

1) $(\Sigma a)(\Pi b)\{(x1Ky2)A(x2Ky3)A(x3Ky4)\}$,

2) $(\Pi a)(\Pi b)\{x1Ky1\}$,

3) $(\Pi a)(\Sigma b)\{(x1Ky3)A(x2Ky4)\}$,

4) $N\{(1)A(2)A(3)\}$, i.e., in all cases besides the ones indicated in the first three lines.

In this form, the connection of many-valued quantifiers with two-valued ones and with two-valued logic in general is obvious.

Such a type of generalization covers two-valued logic as a particular case also on the following level, which we will explain by an example. Suppose $F(a, b, x, y)$ is defined thus:

1) $(\Pi a)(\Pi b)x1Ky1$

2) in all other cases.

Here Π will bind several variables.

On the other hand, this type of generalization covers also the case of a single subject variable. Take, for instance, $F\{a, P(a)\}$ and let it be defined thus: $F\{a, P(a)\} = i$ $(1 \leqslant i \leqslant n)$, if and only if there exists such an a that $P(a) = i$, and for every a, $P(a) = k$ where $k \leqslant i$. By the method of partial normal forms the definition can be written as follows (for simplification of the notation let $P(a) = y)$:

1) $\{(\Sigma a)y1\} K\{(\Pi a)y1\}$,
2) $\{(\Sigma a)y2\} K\{(\Pi a)(y1Ay2)\}$,

....

n) $\{(\Sigma a)yn\} K\{(\Pi a)(y1Ay2A\ldots Ayn)\}$.

In this way, the generalization of the classical quantifiers does not mean the invention of new quantifiers, existing alongside with them, nor the liquidation of the principles of the type 'either all..., or not all...', 'either there exists..., or there does not exist...', etc., of the obvious, customary principles, but it means the generalization of the special functions of the quantifiers so that they can bind any number of variables (in particular: also a single one) and that thus symbols are introduced which can, by binding the variables, transform the proposition into n-valued propositions (into such propositions which can take one of n truth values).

Some words about the negation of quantifiers: suppose x and y differ only by the fact that x contains some quantifier and y its negation, and that the truth value of x and y depends only on the quantifier and its negation (i.e., we take the simplest case). Under this assumption y can be considered as a truth function of x, and the statements of the propositional calculus can be extended to their mutual relations (to the mutual relations between the quantifier and its negation). In two-valued logic the following principles are valid:

1) 'either $\Pi\ldots(x)$, or $N\Pi\ldots(y)$', 'either $\Sigma\ldots(x)$, or $N\Sigma\ldots(y)$;

2) 'if one of x and y is true (false), then the other is false (true)' and based on this the law Axy ($y = Nx$).

How will it be with many-valued quantifiers? Suppose F is a many-valued quantifier. If N is an undetermined negation (meaning 'the state of affairs is not such as it is described by x'), then the more general principle 'either $F\ldots(x)$, or $NF\ldots(y)$' will be valid. As for the second point, different definitions of N are here possible. In particular, N can be defined by a table like the Heyting matrix. Then by the definition of A and N, Axy will not be a law. Or, take still another simple example. Let c be an implication in the sense of Łukasiewicz, $x = P(a, b)$, $y = Q(a, b)$. We define $F(a, b, x, y)$ thus:

1) $F(a, b, x, y) = 1$, if $Cxy = 1$ for all a and for all b;

2) $F(a, b, x, y) = 2$, if there exists such an a that for all b, $Cxy = 2$;

3) $F(a, b, x, y) = 3$ in all other cases.

It can easily be seen that here, and in the examples considered above, F appears (by definition) as an abbreviated designation of the truth conditions of some proposition containing variables. And adding F to the proposition, we transform it (according to the definition) into a proposition which is determinate with respect to the truth value. And only because F binds variables and is defined by means of the quantifiers 'all' and 'there exists', is it considered as a quantifier.

In two-valued logic such an abbreviation is not necessary, since for a single variable and for two truth values an F^1 coincides with Π and an F^2 with Σ. If, e.g., F^1 is defined as follows:

1) $F^1\{a, P(a)\} = 1$, if for all a, $P(a) = 1$;

2) $F^1\{a, P(a)\} = 2$, if there exists an a such that $P(a) = 2$ (or: if not for every a, $P(a) = 1$);

then we actually 'define' a universal quantifier (here a tautological definition is used). But it is sufficient to take two or more variables, or two or more truth values, and the introduction, by definition, of particular symbols F having the properties of quantifiers, becomes possible. But these symbols have a meaning only through the definitions. The evidence which characterizes the two-valued quantifiers is here lacking.

In the case of two or more variables, for two-valued logic the introduction of F's binding several variables is also possible. Take the proposition $a>b$. An F binding simultaneously a and b can easily be introduced: let $F\{a, b, (a>b)\}$ be true, if $(\Pi a)(\Pi b)(a>b)$, and false, if $(\Sigma a)(\Sigma b)(a \leqslant b)$. But as we see, nothing essentially new is introduced here. The point lies, obviously, in the many-valued character of the propositions $P(a)$, $P(a,b)$, $P(a,b,c)$, etc., and from a philosophical point of view, in the occurrence of reasonable interpretations for the truth values in the second of the senses considered above, i.e., in the occurrence of definitions of truth values of propositions which take into account the structure of the propositions. As yet, the question whether this proves useful for the practical aims of science, remains open.

§ 6. In the previous examples subject variables were considered. But predicates can also be covered by a generalized quantification, if a more general argumentation is set up. The examples were exclusively introduced for the purpose of illustrating the very possibility of introducing generalized quantifiers. But the problem is included in the one of introducing such quantifiers with whose help a many-valued generalization of the two-valued predicate calculus could be constructed (as has been done, e.g., in ROSSER and TURQUETTE). Also, the many-valued predicate calculus does not conflict with the two-valued one: it is constructed with the means of two-valued logic itself, and in its content it does not contradict the latter.

In the examples given above, functions of the logic of propositions (*C* and *K*) occurred in the defining part of the definitions of the many-valued quantifier. Thus these functions took only two truth values. Is it in this case possible to exclude the first presupposition, i.e. the many-valued character of the propositions $P(a,b,\ldots)$ and $Q(a,b,\ldots)$ from the number of those necessary for the generalization of quantification? The point is that predicate logic is built upon the logic of propositions; therefore, although Cxy, Kxy and other functions can in the definitions only take two truth values, this does not exclude their many-valued character, as they are functions of the corresponding many-valued logic of propositions. Thus the first presupposition of a generalized quantification (the admission that the propositions can take n truth values dependent on the values of the variables) is of crucial importance.

As has been noted, one of the tasks of the generalized quantifiers consists in making the binding of more than one variable possible. One must add to this that the generalized quantifiers can, in the process of quantification, bind more than one proposition (they can refer to more than one proposition). This can be seen from the examples given.

In the definitions of the type considered, the generalized quantifiers are defined by means of the classical ones. But in principle another way is also possible: first of all to introduce the generalization of the classical quantifiers only along the line of the number of truth values of the propositions, and then, with the help of quantifiers generalized in such a manner to define quantifiers binding more than one variable.

Suppose x is a proposition with one single variable a. The classical quantifiers can, informally, be interpreted as follows:

1) $(\Pi a)x = K(x^1, x^2, \ldots)$, where $x^1, x^2, \ldots$ are propositions obtained by substituting for a all its possible values, and K is two-valued conjunction;

2) $(\Sigma a)x = A(x^1, x^2, \ldots)$, where A is two-valued disjunction.

If 1 is truth and 2 is falsity, then $K(x^1, x^2, \ldots) = max(x^1, x^2, \ldots)$ and $A(x^1, x^2, \ldots) = min(x^1, x^2, \ldots)$. As far as the set of values of a is divided into two subsets, one of which corresponds to truth and the other to falsity, the propositions $(\Pi a)x$ and $(\Sigma a)x$ are two-valued.

But the set of values of a can also be divided into n subsets corresponding to the truth values $1, \ldots, n$ $(n \geqslant 2)$. In such a case, preserving the definitions of K and A, we get many-valued Π and Σ: the propositions $(\Pi a)x$ and $(\Sigma a)x$ take truth values out of the number $1, \ldots, n$ depending on the classification of the values of a.

Between the classical quantifiers and their negations certain relations hold, e.g., the following: if $(\Pi a)x = 1$, then $(\Sigma a)x = 1$; if $(\Pi a)x = 2$, then $(\Sigma a)x = 1$ or $(\Sigma a)x = 2$; if $(\Pi a)x = 1$, then $(N\Pi a)x = 2$; if $(\Pi a)x = 2$, then $(N\Pi a)x = 1$, etc. Similar and, at the same time, generalized relations can also be set up between many-valued Π and Σ, e.g.: if $(\Pi a)x = 1$, then $(\Sigma a)x = 1$; if $(\Pi a)x = 2$, then $(\Sigma a)x = 1$ or $(\Sigma a)x = 2$; if $(\Pi a)x = 3$, then $(\Sigma a)x = 1$ or $(\Sigma a)x = 2$ or $(\Sigma a)x = 3$; ...; if $(\Pi a)x = n$, then $(\Sigma a)x = 1$ or $(\Sigma a)x = 2$ or ... or $(\Sigma a)x = n$; if $(\Pi a)x = i\,(1 \leqslant i \leqslant n)$, then $(N\Pi a)x = n-i+1$, etc. Now these generalized Π and Σ can occur in the definitions of quantifiers binding more than one variable.

The generalized classical quantifiers can also be expressed by means of the classical quantifiers:

1) $(\Pi^* a)x = i$, if $(\Sigma a)(x = i)$ and $(\Pi a)(x \leqslant i)$;

2) $(\Sigma^* a)x = i$, if $(\Sigma a)(x = i)$ and $(\Pi a)(x \geqslant i)$;

(Π^* and Σ^* being the generalized quantifiers). This can also be done for pairs, triads, etc., of variables. For example, $(\Pi^* a, b)(x = i)$, if $(\Sigma a, b)(x = i)$ and $(\Pi a, b)\,(x \leqslant i)$, where $(\Sigma a,b)$ and $(\Pi a, b)$ mean respectively 'there exists a pair of values a and b' and 'for all pairs of values a and b'.

§ 7. The question of the axiomatization of many-valued predicate logic and in general the question of the development of predicate logic on the basis of many-valued propositional logic is the first and perhaps the most important step toward the development of a many-valued conception of logic. As for the other parts of logic (logic of probability, modal, normative, etc., logic), we have already mentioned them in the chapter on the interpretation of many-valued constructions. But this is not only a question of interpretation. It is first of all a question of building a superstructure on the basis of many-valued propositional logic, of the development of a many-valued conception of logic which takes into account such properties of values which are not covered by the logic of propositions and predicates (cf., e.g., ŁUKASIEWICZ 1951).

We have far from exhausted the set of questions which, in the introduction we called the philosophical questions of many-valued logic. And, above all, we have to acknowledge that we only touched to an insignificant extent the questions of a many-valued conception of logic (the influence of the ideas and the apparatus of many-valued logic on the different parts of logic, their use for the description of the rules of reasoning, etc.). However, what has been said on this account seems sufficient for the recognition of the philosophical significance of the ideas of many-valued logic and of the importance of its constructions.

BIBLIOGRAPHY

ACKERMANN, W.

1956. Die Begründung einer strengen Implikation, *The Journal of Symbolic Logic*, 21, 113-128.

BIRKHOFF, G.

1948. *Teorija struktur*, Izdatel'stvo innostranoj literatury, Moscow, 1952. (= *Lattice Theory*, 2nd edition, New York, 1948).

BIRKHOFF, G. and J. VON NEUMANN

1936. The Logic of Quantum Mechanics, *Annals of Mathematics*, 37, 823-843.

BOČVAR, D. A.

1938. Ob odnom trexznačnom isčislenii i ego primenenii k analizu paradoksov klassičeskogo rasširennogo funkcional'nogo isčislenija, *Matematičeskij sbornik*, 4(46), 287-308.

1943. K voprosu o neprotivorečivosti odnogo trexznačnogo isčislenija, *Matematičeskij sbornik*, 12(54), 353-369.

BROUWER, L. E. J.

1923. Ueber die Bedeutung des Satzes vom ausgeschlossenen Dritten in der Mathematik, insbesondere in der Funktionentheorie, *Journal für die reine und angewandte Mathematik*, 154, 1925, 1-7. (Original in Dutch, 1923).

1925. Intuitionistische Zerlegung mathematischer Grundbegriffe, *Jahresbericht der deutschen Mathematiker-Vereinigung*, 33, 251-256.

DESTOUCHES, J. L.

1942. *Principes fondamentaux de physique théorique*, Paris.

DESTOUCHES-FÉVRIER, P.

1951. *La structure des théories physiques*, Paris.

FREY, G.

1954. Bemerkungen zum Problem der mehrwertigen Logiken, *Proceedings of the Second International Congress of the International Union for the Philosophy of Science in Zurich* 1954, Neuchâtel 1955.

GENTZEN, G.

1934. Untersuchungen über das logische Schliessen, *Mathematische Zeitschrift*, 39, 176-210, 405-431.

GLIVENKO, V. M.

1928. Sur la logique de M. Brouwer, *Académie Royale de Belgique, Bulletin de la classe des sciences*, series 5, Vol. 14, 225-228.

1929. Sur quelques points de la logique de M. Brouwer, *Académie Royale de Belgique, Bulletin de la classe des sciences*, series 5, Vol. 15, 183-188.

GÖDEL, K.

1932. Zum intuitionistischen Aussagenkalkül, *Anzeiger der Akademie der Wissenschaften Wien, mathematisch-naturwissenschaftliche Klasse*, 69, 65-66.

GRENIEWSKI, H.

1955. *Elementy logiki formalnej*, Warsaw.

1957. 2^{n+1} wartości logicznych, *Studia filozoficzne*, No. 2, No. 3.

GÜNTHER, G.
1958. Die aristotelische Logik des Seins und die nicht-aristotelische Logik der Reflexion, *Zeitschrift für philosophische Forschung*, 12.

HEYTING, A.
1930. Die formalen Regeln der intuitionistischen Logik, *Sitzungsberichte der Preussischen Akademie der Wissenschaften Berlin, physikalisch-mathematische Klasse*, pp. 42-56.
1956. *Intuitionism: An Introduction*, Amsterdam.
1956a. La conception intuitionniste de la logique, *Les études philosophiques*, 2, 226-233.

HILBERT, D., and W. ACKERMANN,
1938. *Osnovy teoretičeskoj logiki*, Izdatel'stvo innostranoj literatury, Moscow, 1947. (= *Grundzüge der theoretischen Logik*, 2nd edition, Berlin, 1938).

JABLONSKIJ, S. V.
1958. Funkcional'nye postroenija v *k*-značnoj logike, *Trudy matematičeskogo instituta imeni V. A. Steklova*, 51, 5-142.

KALINOWSKI, J.
1953. Teoria zdań normatywnych, *Studia logica*, 1, 113-146.

KLEENE, S. C.
1952. *Vvedenie v metamatematiky*, Izdatel'stvo innostranoj literatury, Moscow, 1957. (= *Introduction to Metamathematics*, Amsterdam, Groningen, New York, 1952).

KOLMOGOROV, A. N.
1925. O principe tertium non datur, *Matematičeskij sbornik*, 32, 646-667.
1932. Zur Deutung der intuitionistischen Logik, *Mathematische Zeitschrift*, 35, pp. 58-65.

KOTARBIŃSKI, T.
1957. *Wykłady z dziejów logiki*, Łódź.

KUZNECOV, B. G.
1959. Ob osnovax kvantovo-reljativistskoj logiki, in: *Logičeskie issledovanija*, Institut Filosofii, Izdatel'stvo Akademii Nauk SSSR, Moscow, pp. 99-112.

LEWIS, C. I., and C. H. LANGFORD
1932. *Symbolic Logic*, New York.

ŁUKASIEWICZ, J.
1920. O pojęciu możliwości, *Ruch filozoficzny*, 5, 169-170.
1920a. O logice trójwartościowej, *Ruch filozoficzny*, 5, 170-171.
1930. Philosophische Bemerkungen zu mehrwertigen Systemen des Aussagenkalküls, *Comptes rendus des séances de la Société des Sciences et des Lettres de Varsovie, classe III*, 23, 51-77.
1951. *Aristotelevskaja sillogistika*, Izdatel'stvo innostranoj literatury, Moscow, 1959. (= *Aristotle's Syllogistic from the Standpoint of Modern Formal Logic*, Oxford, London, New York, 1951).

ŁUKASIEWICZ, J., and A. TARSKI
1930. Untersuchungen über den Aussagenkalkül, *Comptes rendus des séances de la Société des Sciences et des Lettres de Varsovie, classe III*, 23, 30-50.

MOSTOWSKI, A.
1948. *Logika matematyczna*, Wrocław, Warszawa.

NOVIKOV, P. S.
1959. *Élementy matematičeskoj logiki*, Gosudarstvennoe izdatel'stvo fiziko-matematičeskoj literatury, Moscow.

POST, E. L.
1921. Introduction to a General Theory of Elementary Propositions, *American Journal of Mathematics*, 43, 163-185.

RASIOWA, H.
1955. O pewnym fragmencie implikacyjnego rachunku zdań, *Studia logica*, 3, 208-226.

REICHENBACH, H.
1935. *Wahrscheinlichkeitslehre*, Leyden.
1944. *Philosophic Foundations of Quantum Mechanics*, Berkeley, Los Angeles.

ROSSER, J. B. and A. R. TURQUETTE
1952. Many-valued Logics, Amsterdam.

SŁUPECKI, J.
1930. Pełny trójwartościowy rachunek zdań, *Comptes rendus des séances de la Société des Sciences et des Lettres de Varsovie, classe III*, 23.
1936. Der volle dreiwertige Aussagenkalkül, *Comptes rendus des séances de la Société des Sciences et des Lettres de Varsovie, classe III*, 29, 9-11.
1939. Kryterium pełności wielowartościowych systemów logiki zdań, *Comptes rendus des séances de la Société des Sciences et des Lettres de Varsovie, classe III*, 32, 102-109.
1946. Pełny trójwartościowy rachunek zdań, *Annales Universitatis Mariae Curie-Skłodowska*, Lublin, 1, 193-209.

SUSZKO, R.
1957. Formalna teoria wartości logicznych, *Studia logica*, 6.

ŠESTAKOV, V. I.
1946. Predstavlenie xarakteričeskix funkcij predloženij posredstvom vyraženij, realizuemyx relejno-kontaktnymi sxemami, *Izvestija Akademii Nauk, serija matematičeskaja*, 10, 529-554.
1953. Modelirovanie operacij isčislenija predloženij posredstvom prostejšix četyrexpolosnyx sxem, *Vyčislitel'naja matematika i vyčislitel'naja texnika*, 1, 56-89.
1960. O dvojnoj arifmetičeskoj interpretacii trexznačnogo isčislenija vyskazyvanij, *Primenenie logiki v nauke i texnike*, Institut Filosofii, Izdatel'stvo Akademi Nauk SSSR, Moscow, pp. 341-376.

TARSKI, A.
1941. *Vvedenie v logiku i metodologiju deduktivnyx nauk*, Izdatel'stvo innostranoj literatury, Moscow, 1948. (= *Introduction to Logic and to the Methodology of Deductive Sciences*, New York, 1941).

TÖRNEBOHM, H.
1957. On Two Logical Systems Proposed In The Philosophy of Quantum Mechanics, *Theoria*, 3.

WAJSBERG, M.
1931. Aksiomatyzacja trójwartściowego rachunku zdań, *Comptes rendus des séances de la Société des Sciences et des Lettres de Varsovie, classe III*, 24, 126-148.

WEBB, D.
1935. Generation of Any *n*-valued Logic by One Binary Operator, *Proceedings of the National Academy of Sciences*, 21, 252-254.
1936. The Algebra of *n*-valued Logic, *Comptes rendus des séances de la Société des Sciences et des Lettres de Varsovie, classe III*, 29, 153-168.

WRIGHT, G. H. VON
1951. *An Essay in Modal Logic*, Amsterdam.

ZAWIRSKI, Z.
1946. Geneza i rozwój logiki intuicjonistycznej, *Kwartalnik filozoficzny*, 16, 165-222.

ZINOV'EV, A. A.
1958. K problému abstraktního a konkrétního poznatku, *Filosofický Časopis*, 2
1959a. K otázce hodnocení soudů jako abstraktních a konkrétních, *Filosofický Časopis*, 1.
1959b. O logicznej niesprzeczności sądów prawdziwych dotyczących zmian i związków między przedmiotami, *Studia filozoficzne*, 1, 124-133. (German translation in: *Das Widerspruchsprinzip in der neueren sowjetischen Philosophie*, (ed. by N. Lobkowicz), Dordrecht, 1959, pp. 65-76).
1959c. Logičeskoe stroenie znanij o svjazjax, *Logičeskie Issledovanija*, Institut Filosofii, Izdatel'stvo Akademii Nauk SSSR, Moscow, pp. 113-138.
1959d. Sledovanie kak svojstvo vyskazyvanij o svjazjax, *Naučnye Doklady Vysšej Školy, Filosofskie Nauki*, 3, 84-88.
1959e. Problema značenij istinnosti v mnogoznačnoj logike, *Voprosy Filosofii*, 3, pp. 131-136.
1960. *Filosofskie problemy mnogoznačnoj logiki*, Institut Filosofii, Izdatel'stvo Akademii Nauk SSSR. Moscow.
1961a. *Logika vyskazyvanij i teorija vyvoda*, Institut Filosofii, Izdatel'stvo Akademii Nauk SSSR. Moscow.
1961b. Ob odnom sposobe obzora funkcij istinnosti *n*-značnogo isčislenija vyskazyvanij, *Studia logica*, 11, 217f.
1962. Dvuznačnaja i mnogoznačnaja logika, *Filosofskie Voprosy Sovremennoj Formal'noj Logiki*, Institut Filosofii, Izdatel'stvo Akademii Nauk SSSR Moscow, 111-139.

ZINOV'EV, A. A., and I. I. REVZIN
1960. Logičeskaja model' kak sredstvo naučnogo issledovanija, *Voprosy Filosofii*, 1, 82-90.

TRANSLITERATION TABLE

The system used here is identical with that of the Institute of East European Studies at the University of Fribourg

Russian Letter *Cap. Ital.*	Studies in Soviet Thought	U.S. Library of Congress	Russian Letter *Cap. Ital.*	Studies in Soviet Thought	U.S. Library of Congress
А *а*	a	a	Р *р*	r	r
Б *б*	b	b	С *с*	s	s
В *в*	v	v	Т *т*	t	t
Г *г*	g	g	У *у*	u	u
Д *д*	d	d	Ф *ф*	f	f
Е *е*	e	e	Х *х*	x	k͡h
Ё *ё*	ë	ë	Ц *ц*	c	t͡s
Ж *ж*	ž	z͡h	Ч *ч*	č	c͡h
З *з*	z	z	Ш *ш*	š	s͡h
И *и*	i	i	Щ *щ*	šč	s͡hch
Й *й*	j	ĭ	Ъ *ъ*	″	″
К *к*	k	k	Ы *ы*	y	y
Л *л*	l	l	Ь *ь*	′	′
М *м*	m	m	Э *э*	é	é
Н *н*	n	n	Ю *ю*	ju	i͡u
О *о*	o	o	Я *я*	ja	i͡a
П *п*	p	p			

TRANSLATORS' NOTES

1. In Russian, the word 'variable' ('*peremennaja*') is both an adjective and a noun, and there is no way of distinguishing, except by context, whether the Russian expression '*peremennye vyskazyvanija*' means 'variable propositions' or 'propositional variables'. We have translated it as 'propositional variables', since this is the normal expression in English, except in a few cases, where the context indicates that the author wished to emphasize the concept of 'variable propositions'.
2. We translate the Russian word '*vyskazyvanie*' by 'proposition', where 'proposition' has to be understood in the traditional sense meaning a declarative sentence together with its meaning (cf. A. Church, 'Propositions and sentences', p. 3 in J. M. Bocheński, A. Church, N. Goodman *The problem of universals*, Notre Dame

(Indiana) 1956). In contemporary Soviet logic the following three terms are used:
'*predloženie*' as a translation of 'sentence';
'*suždenie*' as a translation of 'proposition', where 'proposition' is understood in the modern, abstract sense;
'*vyskazyvanie*' as a general term covering both '*predloženie*' and '*suždenie*'.
Cf. the remarks of V. A. Uspenskij, editor of the translation of Church's *Introduction to Mathematical Logic* (*Vvedenie v matematičeskuju logiku, tom. 1*, Moscow, 1960, pp. 10-11); Izdatel'stvo innostranoj literatury; the same usage has also been observed in the translation of Carnap's *Meaning and Necessity* (*Značenie i neobxodimost'*, Izdatel'stvo innostranoj literatury, Moscow, 1959). Zinov'ev uses in his book only the term '*vyskazyvanie*'.

3. The Russian word in the text of Zinov'ev is literally 'operator'. According to the terminology of Church, an operator is a symbol which has the function of binding variables, is e.g., a quantifier. The symbol for what Zinov'ev calls 'operator' has in Church the name 'connective' or 'functional constant' (cf. A. Church, *Introduction to Mathematical Logic*, Vol. 1, Princeton 1956, pp. 31, 34, 39).
4. In Polish logical literature '*E*' is the symbol of equivalence and '*J*' the symbol of exclusive disjunction.
5. Of course, one way out is possible: we can introduce into parenthesis-free notation a *new* symbol, a special three-place connective standing for exclusive disjunction of three arguments.
6. This is the traditional name. However, it would be more consequent to call it 'law of non-contradiction'.
7. According to B. Sobociński, 'Jan Łukasiewicz 1878-1956', *Rocznik Polskiego Towarzystwa Naukowego na Obczyźnie*, London, 1956-57, p. 12, Łukasiewicz constructed the first many-valued system in the summer of 1917.
8. A further example is *CNKxyANxNy* (cf. J. M. Bocheński and L. H. Hackstaff 'A study in many-valued logic', *Studies in Soviet Thought*, Vol. 2, 1962, 37-48).
9. This holds for $n > 2$ only.
10. '*predloženie*' has now become the translation of the English 'sentence' (cf. above translator's note 2)). Bočvar's term '*predloženie*' we will translate by the complex expression 'meaningful proposition'.
11. cf. above translator's note 1.
12. According to the terminology of A. Church (cf. *Introduction to Mathematical Logic*, Vol. 1, Princeton 1956, p. 9) the values are not the constants but what the constants stand for.
13. Zinov'ev 's method of using two different symbols (a and α) in the quantifier and in the quantified formula is very unusual. Also, he abandons this kind of notation in what follows. – Furthermore, negation is normally not considered as belonging to the quantifier but rather as negating the whole quantified formula.

SUBJECT INDEX

NAME INDEX